The Essential Calculus Workbook

LIMITS AND DERIVATIVES

Tim Hill

Questing Vole Press

The Essential Calculus Workbook: Limits and Derivatives
by Tim Hill

Editor: Kevin Debenjak
Proofreader: Diane Yee
Compositor: Kim Frees
Cover: Questing Vole Press

Contents

1 The Slope of the Tangent Line

Problem 1.1 *The tangent line* Describe the geometric nature of a tangent line.

Solution Calculus is usually divided into two main parts: **differential calculus** and **integral calculus**, each with its own notation, terminology, and computational methods. The basic problem of differential calculus, the topic of this book, is the problem of tangents: calculate the slope of the tangent line to the graph of $y = f(x)$ at a given point P.

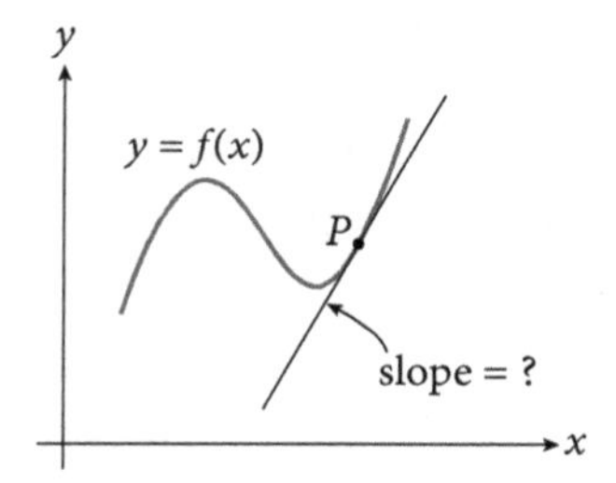

Almost all the ideas and applications of differential calculus revolve around this geometric problem. At first glance this problem appears to be limited in scope, but in fact it has far-reaching and profound real-world applications to the sciences and other fields outside of mathematics and geometry. (The same can be said of the basic problem of integral calculus, the problem of areas: calculate the area under the graph of $y = f(x)$ between the points $x = a$ and $x = b$.)

Before attempting to calculate the slope of a tangent line, we must first decide what a tangent line *is*.

The case of a circle is simple. A tangent to a circle is a line that intersects the circle at only one point, called the **point of tangency**; lines that aren't tangents either intersect the circle at two different points or miss it entirely.

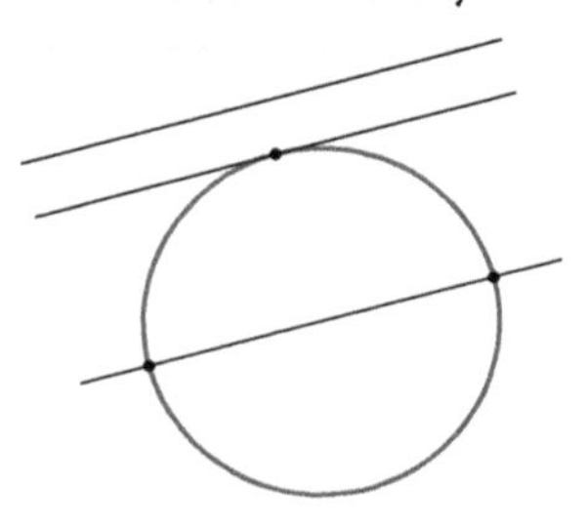

This situation reflects the intuitive idea that most of us have that a tangent to a curve at a given point is a line that "just touches" the curve at that point. (The Latin word *tangere* means "to touch", as in the English word *tangible.*) It also suggests the possibility of defining a tangent to a curve as a line that intersects the curve at only one point. This definition (used successfully by the ancient Greeks) works for circles and a few other special curves but is unsatisfactory for curves in general. To understand why, consider the curve shown in the following figure: it has an acceptable tangent (the lower line) that this definition would reject, and an obvious nontangent (the upper line) that this definition would accept.

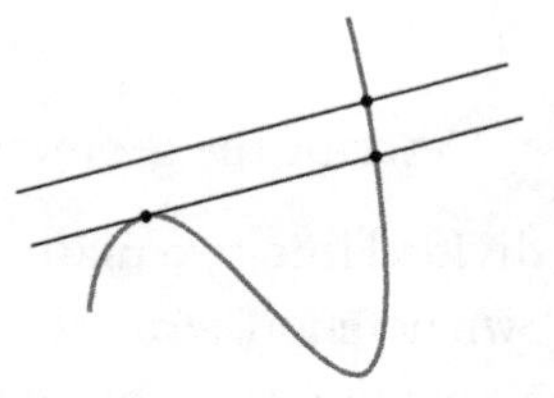

The modern concept of a tangent line, which originated in the 1600s, describes the geometric nature of tangents and provides the key to a practical process for constructing tangents. Consider a curve $y = f(x)$, and let P be a given fixed point on this curve. Let Q be a second nearby point on the curve, and draw the secant line PQ. The tangent line at P can now be thought of as the limiting position of the variable secant as Q slides along the curve toward P.

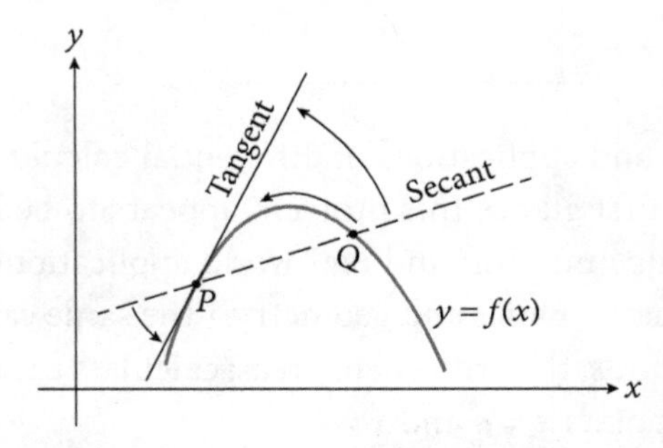

This concept of tangents is one of the most useful ideas in all of mathematics, leading directly to precise descriptions of velocity, acceleration, force, orbits, reflection, refraction, Newtonian dynamics, and other physical phenomena. Problem 1.2 shows how this qualitative idea leads to a quantitative method for calculating the exact slope of the tangent in terms of the given function $f(x)$.

Problem 1.2 *Calculating the slope of the tangent* Calculate the slope of the tangent to the curve $y = x^2$ at the point (x_0, y_0).

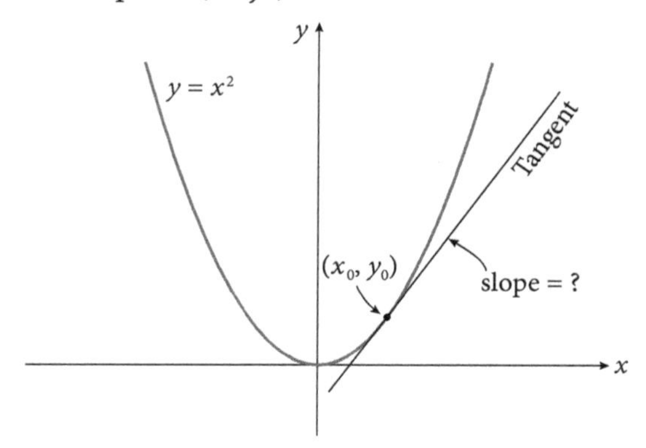

Solution Let $P = (x_0, y_0)$ be an arbitrary fixed point on the parabola $y = x^2$. To start the process of calculating the slope of the tangent to this parabola at the given point P, we choose a second nearby point $Q = (x_1, y_1)$ on the curve.

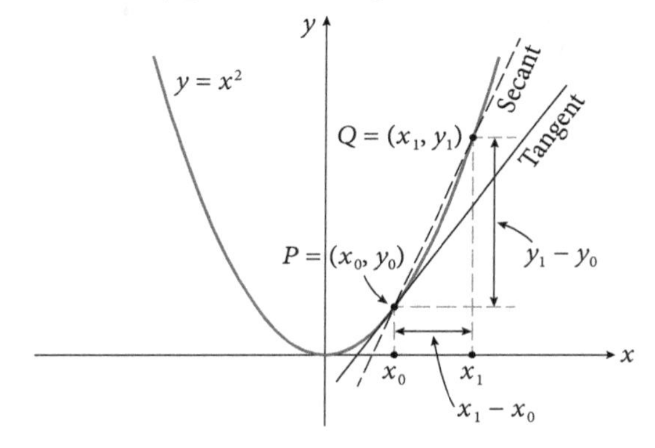

Next, we draw the secant line PQ that's determined by these two points. The slope of this secant is

$$(1) \quad m_{\text{sec}} = \text{slope of } PQ = \frac{y_1 - y_0}{x_1 - x_0}.$$

Now for the crucial step: we let x_1 approach x_0, so that the variable point Q approaches the fixed point P by sliding along the curve much like a bead sliding along a curved wire. As this happens, the secant changes direction and visibly approaches the tangent at P as its limiting position. Also, it's clear intuitively that the slope m of the tangent is the limiting value approached by the slope m_{sec} of the secant. If we use the standard symbol $\rightarrow$ to mean "approaches", then the preceding statement can be expressed concisely as

$$(2) \quad m = \lim_{Q \to P} m_{\text{sec}} = \lim_{x_1 \to x_0} \frac{y_1 - y_0}{x_1 - x_0}.$$

The abbreviation "lim", with "$x_1 \rightarrow x_0$" written below it, is read "the limit, as x_1 approaches x_0, of . . .".

We can't calculate the limiting value m in (2) by simply setting $x_1 = x_0$, because then $y_1 = y_0$ and this would give the meaningless result

$$m = \frac{y_0 - y_0}{x_0 - x_0} = \frac{0}{0}.$$

We must think of x_1 as coming very close to x_0 *but remaining distinct from it.* As this happens, however, both $y_1 - y_0$ and $x_1 - x_0$ become arbitrarily small, and it's unclear what limiting value their quotient approaches.

The way out of this difficulty is to use the equation of the curve. Because P and Q both lie on the curve, we have $y_0 = x_0^2$ and $y_1 = x_1^2$, so (1) can be written

$$(3) \quad m_{\text{sec}} = \frac{y_1 - y_0}{x_1 - x_0} = \frac{x_1^2 - x_0^2}{x_1 - x_0}.$$

The reason that this numerator becomes small is that it contains the denominator $x_1 - x_0$ as a factor. If this common factor is canceled, then we obtain

$$m_{\text{sec}} = \frac{y_1 - y_0}{x_1 - x_0} = \frac{x_1^2 - x_0^2}{x_1 - x_0} = \frac{(x_1 - x_0)(x_1 + x_0)}{x_1 - x_0} = x_1 + x_0.$$

and (2) becomes

$$m = \lim_{x_1 \to x_0} \frac{y_1 - y_0}{x_1 - x_0} = \lim_{x_1 \to x_0} (x_1 + x_0).$$

It's now easy to see what's happening: as x_1 gets closer and closer to x_0, $x_1 + x_0$ becomes more and more nearly equal to $x_0 + x_0 = 2x_0$. Accordingly,

$$(4) \quad m = 2x_0$$

is the slope of the tangent to the curve $y = x^2$ at the point (x_0, y_0).

Problem 1.3 Compute the slopes of the tangents at the points (1, 1) and (−½, ¼) on the parabola $y = x^2$.

Solution By using formula (4) in Problem 1.2, we have slopes $m = 2 \cdot 1 = 2$ at (1, 1) and $m = 2 \cdot -½ = -1$ at (−½, ¼).

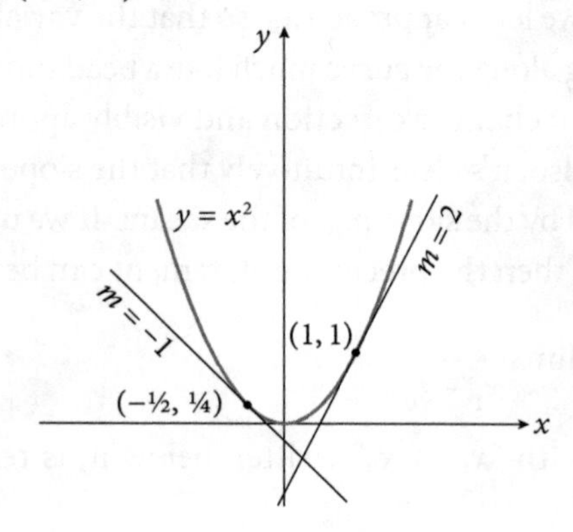

Using the point-slope form of the equation of a line, our two tangent lines have equations

$$\frac{y-1}{x-1}=2 \quad \text{and} \quad \frac{y-\frac{1}{4}}{x+\frac{1}{2}}=-1.$$

In exactly the same way,

$$\frac{y-x_0^{\ 2}}{x-x_0}=2x_0$$

is the equation of the tangent at a general point (x_0, x_0^2) on the curve.

Problem 1.4 *Delta notation* Reformulate the calculations in Problem 1.2 by using **delta notation** $\Delta x = x_1 - x_0$ in formula (3) and then generalize the result for any function.

Solution The procedure begins by changing the independent variable x from a first value x_0 to a second value x_1. The standard notation for the amount of such a change is Δx (read "delta x"), so that

$$(5) \quad \Delta x = x_1 - x_0$$

is the change in x in going from the first value to the second. We can also think of the second value as being obtained from the first by adding the change:

$$(6) \quad x_1 = x_0 + \Delta x.$$

Note that Δx isn't the product of a number Δ and a number x; it's a single number called an **increment** of x. An increment Δx can be either positive or negative. If $x_0 = 1$ and $x_1 = 3$, for example, then $\Delta x = 3 - 1 = 2$; and if $x_0 = 1$ and $x_1 = -2$, then $\Delta x = -2 - 1 = -3$.

The letter Δ is the Greek *d*; when it's written in front of a variable, it denotes the difference between two values of that variable. This simple notation is so convenient that it has spread to almost every part of mathematics and science. We illustrate its role here by using it to reformulate the calculations in Problem 1.2.

In view of (5) and (6), formula (3) for the slope of the secant can be written in the form

$$(7) \quad m_{\text{sec}} = \frac{x_1^{\ 2} - x_0^{\ 2}}{x_1 - x_0} = \frac{(x_0+\Delta x)^2 - x_0^{\ 2}}{\Delta x}.$$

This time, instead of factoring the numerator, we expand its first term and simplify the result, obtaining

$$\begin{aligned}(x_0+\Delta x)^2 - x_0^{\ 2} &= x_0^{\ 2} + 2x_0\Delta x + (\Delta x)^2 - x_0^{\ 2} \\ &= 2x_0\Delta x + (\Delta x)^2 \\ &= \Delta x(2x_0 + \Delta x),\end{aligned}$$

so (7) becomes

$$m_{\text{sec}} = 2x_0 + \Delta x.$$

If we insert this in (2) and use the fact that $x_1 \to x_0$ is equivalent to $\Delta \to 0$, then

$$m = \lim_{\Delta x \to 0} (2x_0 + \Delta x) = 2x_0,$$

as before. Again it's easy to see what's happening in this limit process: as Δx gets closer and closer to zero, $2x_0 + \Delta x$ becomes more and more nearly equal to $2x_0$.

This second method, using delta notation, depends on expanding the square $(x_0 + \Delta x)^2$, whereas the first depends on factoring the expression $x_1^2 - x_0^2$. In this particular case neither calculation is noticeably harder than the other. In general, however, expanding is easier than factoring, and for this reason the method of increments is standard procedure.

The calculation that we've just carried out for the parabola $y = x^2$ can be described in principle for the graph of any function $y = f(x)$.

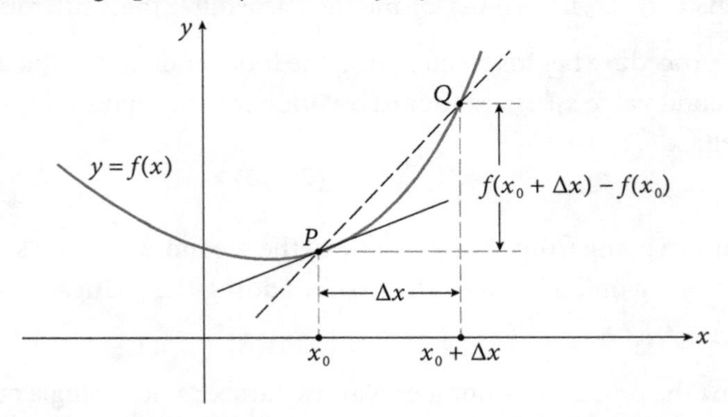

We first compute the slope of the secant through the two points P and Q corresponding to x_0 and $x_0 + \Delta x$,

$$m_{\text{sec}} = \frac{f(x_0 + \Delta x) - f(x_0)}{\Delta x}.$$

We then calculate the limit of m_{sec} as Δx approaches zero, obtaining a number m that we interpret geometrically as the slope of the tangent to the curve at the point P:

$$m = \lim_{\Delta x \to 0} \frac{f(x_0 + \Delta x) - f(x_0)}{\Delta x}.$$

The value of this limit is usually denoted by the symbol $f'(x_0)$, read "f prime of x_0", to emphasize its dependence on both the point x_0 and the function $f(x)$. Thus, by definition we have

$$(8) \quad f'(x_0) = \lim_{\Delta x \to 0} \frac{f(x_0 + \Delta x) - f(x_0)}{\Delta x}.$$

In this notation, the result of the calculation given above can be expressed as follows: if $f(x) = x^2$, then $f'(x_0) = 2x_0$.

Problem 1.5 Calculate $f'(x_0)$ if $f(x) = 2x^2 - 3x$.

Solution For this function, the numerator of the quotient in (8) in Problem 1.4 is

$$\begin{aligned}
f(x_0+\Delta x)-f(x_0) &= [2(x_0+\Delta x)^2-3(x_0+\Delta x)]-[2x_0{}^2-3x_0] \\
&= 2x_0{}^2+4x_0\Delta x+2(\Delta x)^2-3x_0-3\Delta x-2x_0{}^2+3x_0 \\
&= 4x_0\Delta x+2(\Delta x)^2-3\Delta x \\
&= \Delta x(4x_0+2\Delta x-3).
\end{aligned}$$

The quotient in (8) is therefore

$$\frac{f(x_0+\Delta x)-f(x_0)}{\Delta x}=4x_0+2\Delta x-3,$$

and

$$\begin{aligned}
f'(x_0) &= \lim_{\Delta x\to 0}(4x_0+2\Delta x-3) \\
&= 4x_0-3.
\end{aligned}$$

Problem 1.6 For the limit to exist in (8) in Problem 1.4, show that it's necessary to have the same limiting value for both directions of approach (that is, from the right and the left).

Solution We have assumed in the procedure leading to (8) that the curve under discussion actually has a single definite tangent at the point *P*. But some curves don't have such a tangent at every point.

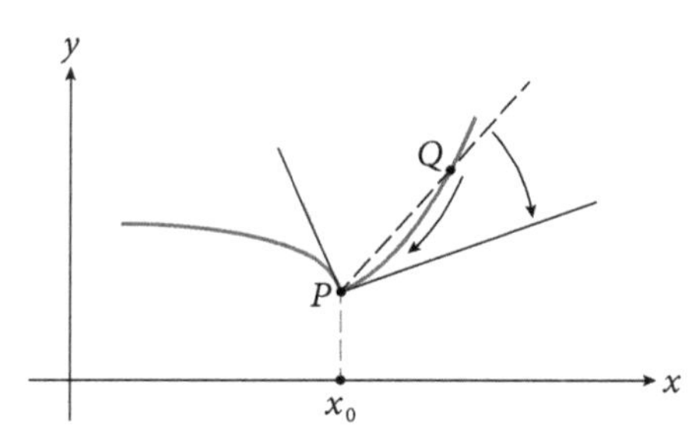

When a tangent exists, however, it's clearly necessary for the secant *PQ* to approach the same limiting position whether *Q* approaches *P* from the right or from the left. These two modes of approach correspond, respectively, to Δx approaching zero through only positive or only negative values. It's therefore part of the meaning of (8) that for this limit to exist we must have the same limiting value for both directions of approach.

Problem 1.7 Find the equation of the tangent to the parabola $y = x^2$

(a) at the point (–2, 4)

(b) at the point where the slope is 8

(c) if the x-intercept of the tangent is 2

Solution (a) Use the last equation in Problem 1.3: $x_0 = -2$, so the tangent line is $(y - 4)/(x + 2) = -4$; that is, $4x + y + 4 = 0$.

(b) The slope at (x_0, x_0^2) is $2x_0$, so $2x_0 = 8$; that is, $8x - y - 16 = 0$.

(c) If the tangent meets the parabola at (x_0, x_0^2), then its equation is $(y - x_0^2)/(x - x_0) = 2x_0$. Because $(2, 0)$ is on the tangent, $(0 - x_0^2)/(2 - x_0) = 2x_0$; that is, $x_0(4 - x_0) = 0$. So x_0 is either 0 or 4, and the line is either $(y - 0)/(x - 0) = 0$ or $(y - 16)/(x - 4) = 8$; that is, either $y = 0$ or $8x - y - 16 = 0$. Because $y = 0$ does not have a unique x-intercept, the equation is $8x - y - 16 = 0$.

Problem 1.8 A straight line $y = mx + b$ is its own tangent line at any point. Verify this fact by using formula (8) in Problem 1.4 to show that $f'(x_0) = m$ if $f(x) = mx + b$.

Solution

$$f'(x_0) = \lim_{\Delta x \to 0} \frac{[m(x_0 + \Delta x) + b] - [mx_0 + b]}{\Delta x}$$

$$= \lim_{\Delta x \to 0} \frac{m\Delta x}{\Delta x} = \lim_{\Delta x \to 0} m = m.$$

Problem 1.9 Use formula (8) in Problem 1.4 to calculate $f'(x_0)$ if $f(x)$ is equal to

(a) $x^2 - 4x - 5$
(b) $x^2 - 2x + 1$
(c) $2x^2 + 1$
(d) $x^2 - 4$

Solution

(a) $$f'(x_0) = \lim_{\Delta x \to 0} \frac{[(x_0 + \Delta x)^2 - 4(x_0 + \Delta x) - 5] - [x_0^2 - 4x_0 - 5]}{\Delta x}$$
$$= \lim_{\Delta x \to 0} (2x_0 + \Delta x - 4)$$
$$= 2x_0 - 4.$$

(b) $$f'(x_0) = \lim_{\Delta x \to 0} \frac{[(x_0 + \Delta x)^2 - 2(x_0 + \Delta x) + 1] - [x_0^2 - 2x_0 + 1]}{\Delta x}$$
$$= \lim_{\Delta x \to 0} (2x_0 + \Delta x - 2)$$
$$= 2x_0 - 2.$$

(c) $$f'(x_0) = \lim_{\Delta x \to 0} \frac{[2(x_0 + \Delta x)^2 + 1] - [2x_0^2 + 1]}{\Delta x}$$
$$= \lim_{\Delta x \to 0} (4x_0 + 2\Delta x)$$
$$= 4x_0.$$

(d) $$f'(x_0) = \lim_{\Delta x \to 0} \frac{[(x_0 + \Delta x)^2 - 4] - [x_0^2 - 4]}{\Delta x}$$
$$= \lim_{\Delta x \to 0} (2x_0 + \Delta x)$$
$$= 2x_0.$$

Problem 1.10 Find the equation of the tangent line to the curve $y = 2x^2 + 1$ that is parallel to the line $8x + y - 2 = 0$. Hint: Use Problem 1.9(c).

Solution The slope is -8, so, by Problem 1.9(c), we must solve $4x_0 = -8$. We get $x_0 = -2$, so $f(x_0) = 9$ and the tangent line is $(y - 9)/(x + 2) = -8$; that is, $8x + y + 7 = 0$.

Problem 1.11 Prove analytically (that is, without appealing to geometric reasoning) that there is no line through the point $(1, -2)$ that is tangent to the curve $y = x^2 - 4$. Hint: Use Problem 1.9(d).

Solution By Problem 1.9(d), the equation of the tangent at $(x_0, x_0 - 4)$ is

$$\frac{y - (x_0^2 - 4)}{x - x_0} = 2x_0;$$

that is, $y + x_0^2 + 4 = 2x_0 x$. If this line passes through $(1, -2)$, then $x_0^2 - 2x_0 + 2 = 0$. But $x_0^2 - 2x_0 + 2 = (x_0 - 1)^2 + 1 \geq 1$, so no line through $(1, -2)$ is tangent to the curve.

Problem 1.12 Find equations for the two lines through the point $(3, 13)$ that are tangent to the parabola $y = 6x - x^2$.

Solution Let $(a, 6a - a^2)$ be the point of tangency. We first find the slope of this tangent via equation (8) in Problem 1.4 with $f(x) = 6x - x^2$ and $x_0 = a$; that is,

$$f'(x_0) = \lim_{\Delta x \to 0} \frac{[6(x_0 + \Delta x) - (x_0 + \Delta x)^2] - [6x_0 - x_0^2]}{\Delta x}$$
$$= \lim_{\Delta x \to 0} (6 - 2x_0 - \Delta x)$$
$$= 6 - 2x_0.$$

Then the equation of the tangent is

$$\frac{y - (6a - a^2)}{x - a} = 6 - 2a,$$

or $(6 - 2a)x - y + a^2 = 0$. If $(3, 13)$ lies on this tangent, then $(6 - 2a) \cdot 3 - 13 + a^2 = a^2 - 6a + 5 = (a - 5)(a - 1) = 0$. Hence $a = 1$ or $a = 5$, and the equations for the two tangents are $4x - y + 1 = 0$ and $-4x - y + 25 = 0$.

Problem 1.13 For what value of b does the graph of $y = x^2 + bx + 1$ have a horizontal tangent at $x = 3$?

Solution Because $y' = 2x + b$, the slope of the tangent at $x = 3$ is $2 \cdot 3 + b = b + 6$, so $b = -6$.

Problem 1.14 Let $P = (x_0, y_0)$ be a point on the parabola $y = x^2$. Show that a nonvertical line passing through P which does not intersect the curve at any other point is necessarily the tangent at P; that is, show that if the line

$$y - y_0 = m(x - x_0)$$

intersects $y = x^2$ only at (x_0, y_0), then $m = 2x_0$.

Solution We must solve $y - y_0 = m(x - x_0)$ and $y = x^2$. Using $y_0 = x_0^2$, this becomes $x^2 - x_0^2 = m(x - x_0)$, or $(x - x_0)(x + x_0 - m) = 0$. Hence the line and the parabola intersect at $(m - x_0, (m - x_0)^2)$, which is distinct from (x_0, x_0^2) if $m \neq 2x_0$.

---

2 The Definition of the Derivative

Problem 2.1 *Definition of the derivative* Use formula (8) in Problem 1.4 to define the derivative of any function $f(x)$.

Solution If we separate formula (8) from its geometric motivation and drop the subscript on x_0, then we arrive at our basic definition: given any function $f(x)$, its **derivative** $f'(x)$ is the new function whose value at a point x is defined by

$$(1) \quad f'(x) = \lim_{\Delta x \to 0} \frac{f(x+\Delta x) - f(x)}{\Delta x}.$$

In calculating this limit, x is held fixed while Δx varies and approaches zero. The limit might exist for some values of x and fail to exist for other values. If the limit exists for $x = a$, then the function is said to be **differentiable** at a. A **differentiable function** is one that is differentiable at each point of its domain. Most of the specific functions considered in this book have this property.

Problem 2.2 *Geometric interpretation of the derivative* Give a geometric interpretation of the derivative of $f(x)$ and describe the limits of such an interpretation.

Solution The derivative $f'(x)$ can be visualized in the way illustrated by the following figure, in which $f(x)$ is the variable height of a point P moving along the curve and $f'(x)$ is the variable slope of the tangent at P.

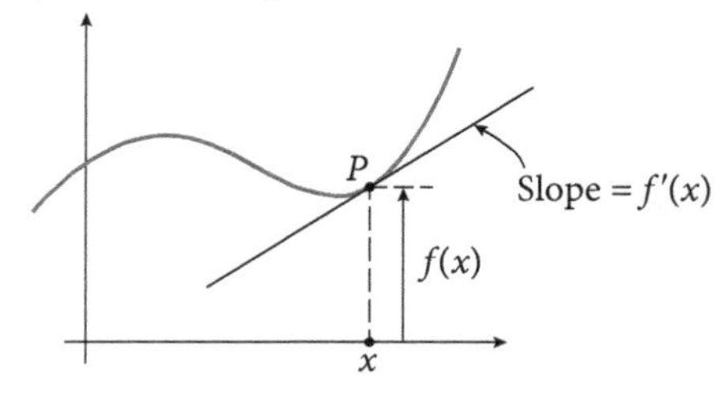

Strictly speaking, however, the definition (1) of the derivative doesn't depend in any way on geometric ideas. The preceding figure is a **geometric interpretation**, and important as it may be as an aid to understanding, it's not an essential part of the concept of the derivative. In the next chapter we'll meet other equally important interpretations that have nothing to do with geometry. We must therefore consider $f'(x)$ purely as a function, and recognize that it has several interpretations but no necessary connection with any one of them.

Problem 2.3 *Differentiation and the three-step rule* Describe the differentiation of a given function $f(x)$ as a systematic procedure.

Solution The process of actually forming the derivative $f'(x)$ is called the **differentiation** of the given function $f(x)$. This process is the fundamental operation of calculus, on which everything else depends. In principle, we follow the computational instructions specified in (1) in Problem 2.1. These instructions can be arranged into a systematic procedure called the **three-step rule.**

STEP 1 Form the difference $f(x + \Delta x) - f(x)$ for the given function, simplifying it if possible to the point where Δx is a factor.

STEP 2 Divide by Δx to form the **difference quotient**

$$\frac{f(x+\Delta x)-f(x)}{\Delta x},$$

and manipulate this expression to prepare the way for evaluating its limit as $\Delta x \to 0$. In most of the problems in this book, this manipulation involves nothing more than canceling Δx from the numerator and denominator.

STEP 3 Evaluate the limit of the difference quotient as $\Delta x \to 0$. If Step 2 has accomplished its purpose, then a simple inspection is usually all that's needed here.

Keep in mind that the simple-looking notation $f(x)$ encompasses all conceivable functions, so these steps are sometimes easy to carry out and sometimes quite difficult. Even examples that depend on only elementary algebra can require a little ingenuity.

Problem 2.4 Find $f'(x)$ if $f(x) = x^3$.

Solution

STEP 1:

$$\begin{aligned} f(x+\Delta x)-f(x) &= (x+\Delta x)^3 - x^3 \\ &= x^3 + 3x^2\Delta x + 3x(\Delta x)^2 + (\Delta x)^3 - x^3 \\ &= 3x^2\Delta x + 3x(\Delta x)^2 + (\Delta x)^3 \\ &= \Delta x[3x^2 + 3x\Delta x + (\Delta x)^2]. \end{aligned}$$

STEP 2:

$$\frac{f(x+\Delta x)-f(x)}{\Delta x} = 3x^2 + 3x\Delta x + (\Delta x)^2.$$

STEP 3:

$$f'(x) = \lim_{\Delta x \to 0} [3x^2 + 3x\Delta x + (\Delta x)^2] = 3x^2.$$

Problem 2.5 Find $f'(x)$ if $f(x) = 1/x$.

Solution

STEP 1:

$$f(x+\Delta x) - f(x) = \frac{1}{x+\Delta x} - \frac{1}{x}$$
$$= \frac{x-(x+\Delta x)}{x(x+\Delta x)}$$
$$= \frac{-\Delta x}{x(x+\Delta x)}.$$

STEP 2:

$$\frac{f(x+\Delta x) - f(x)}{\Delta x} = \frac{-1}{x(x+\Delta x)}.$$

STEP 3:

$$f'(x) = \lim_{\Delta x \to 0} \frac{-1}{x(x+\Delta x)} = -\frac{1}{x^2}.$$

Problem 2.6 What does the result of Problem 2.5 tell us about the graph of the function $y = f(x) = 1/x$?

Solution First, $f'(x) = -1/x^2$ is clearly negative for all $x \neq 0$, and because this is the slope of the tangent, all tangent lines point down to the right. Also, when x is near 0, $f'(x)$ is very large, which means that these tangent lines are steep; and when x is large, $f'(x)$ is small, so these tangent lines are nearly horizontal. We can verify these observations by examining the following figure. (In general, derivatives reveal much about the behavior of functions and the properties of their graphs because the derivative at a point gives the slope of the tangent at that point.)

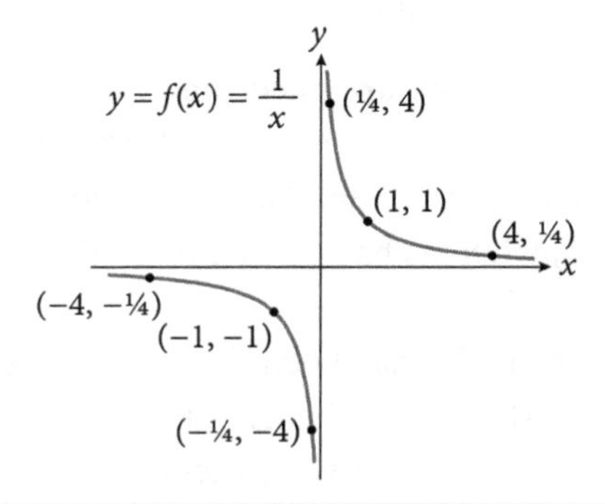

Problem 2.7 Find $f'(x)$ if $f(x) = \sqrt{x}$.

Solution

STEP 1:

$$f(x+\Delta x)-f(x)=\sqrt{x+\Delta x}-\sqrt{x}.$$

STEP 2:

$$\frac{f(x+\Delta x)-f(x)}{\Delta x}=\frac{\sqrt{x+\Delta x}-\sqrt{x}}{\Delta x}.$$

This form isn't convenient for canceling the Δx's, so we use an algebraic trick to remove the square roots from the numerator. Multiply both numerator and denominator of the last fraction by

$$\sqrt{x+\Delta x}+\sqrt{x},$$

which amounts to multiplying this fraction by 1, and then simplify the numerator by using the algebraic identity $(a - b)(a + b) = a^2 - b^2$:

$$\begin{aligned}\frac{f(x+\Delta x)-f(x)}{\Delta x}&=\frac{\sqrt{x+\Delta x}-\sqrt{x}}{\Delta x}\cdot\frac{\sqrt{x+\Delta x}+\sqrt{x}}{\sqrt{x+\Delta x}+\sqrt{x}}\\&=\frac{(x+\Delta x)-x}{\Delta x\left(\sqrt{x+\Delta x}+\sqrt{x}\right)}\\&=\frac{1}{\sqrt{x+\Delta x}+\sqrt{x}}.\end{aligned}$$

The next step is now easy.

STEP 3:

$$f'(x)=\lim_{\Delta x\to 0}\frac{1}{\sqrt{x+\Delta x}+\sqrt{x}}=\frac{1}{\sqrt{x}+\sqrt{x}}=\frac{1}{2\sqrt{x}}.$$

Problem 2.8 *Leibniz notation* Compare prime notation $f'(x)$ to Leibniz notation dy/dx for differentiation, noting situations where one is superior to the other. (Prime notation is also called Lagrange notation.)

Solution Several different notations are in common use for derivatives, with preference varying according to the circumstances in which the symbols are being used. The choice of notation matters a great deal. Good notations can smooth the way and do much of our work for us, whereas bad ones muddy the mathematical waters and hinder progress.

The derivative of a function $f(x)$ has been denoted above by $f'(x)$. This notation has the merit of emphasizing that the derivative of $f(x)$ is another function of x that's associated in a particular way with the given function. If our function is given in the form $y = f(x)$, with the dependent variable displayed, then the shorter symbol y' is often used in place of $f'(x)$.

The main disadvantage of this prime notation for derivatives is that it doesn't suggest the nature of the process by which $f'(x)$ is obtained from $f(x)$. The notation devised by Gottfried Wilhelm Leibniz (who conceived calculus independently of Isaac Newton) for his version of calculus is better in this respect, and in other ways as well.

To explain Leibniz's notation, we begin with a function $y = f(x)$ and write the difference quotient

$$\frac{f(x+\Delta x)-f(x)}{\Delta x}$$

in the form

$$\frac{\Delta y}{\Delta x},$$

where $\Delta y = f(x + \Delta x) - f(x)$. Here Δy isn't just any change in y; it's the specific change that results when the independent variable is changed from x to $x + \Delta x$. As we know, the difference quotient $\Delta y/\Delta x$ can be interpreted as the ratio of the change in y to the change in x along the curve $y = f(x)$, and this ratio is the slope of the secant, as shown in the following figure.

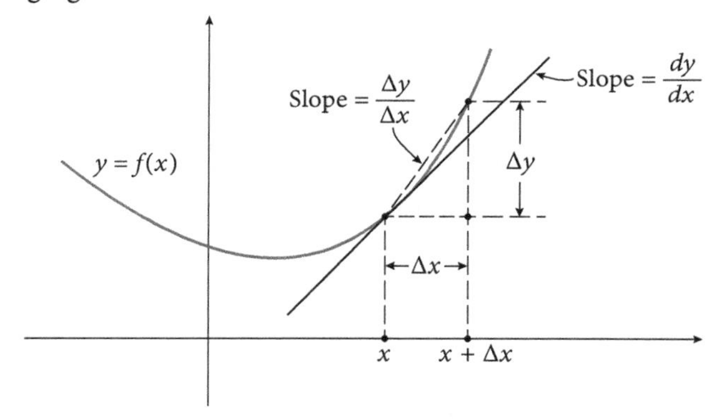

Leibniz wrote the limit of this difference quotient, which of course is the derivative $f'(x)$, in the form dy/dx (read "*dy* over *dx*" or "*dy dx*"). In this notation, the definition of the derivative becomes

$$\text{(2)} \quad \frac{dy}{dx} = \lim_{\Delta x \to 0} \frac{\Delta y}{\Delta x},$$

and this is the slope of the tangent in the preceding figure. Two slightly different equivalent forms of dy/dx are

$$\frac{df(x)}{dx} \quad \text{and} \quad \frac{d}{dx}f(x).$$

In the second of these, the symbol d/dx should be thought of as an operation that can be applied to the function $f(x)$ to yield its derivative $f'(x)$, as denoted by the equation

$$\frac{d}{dx}f(x) = f'(x).$$

The symbol d/dx can be read "the derivative with respect to x of ..." whatever function of x follows it.

It's crucial to understand that dy/dx in (2) is a single indivisible symbol. In spite of the way it's written, it's *not* the quotient of two quantities dy and dx because dy and dx haven't been defined and have no independent existence. In Leibniz notation, the formation of the limit on the right of (2) is symbolically expressed by replacing the letter Δ by the letter d. From this point of view, the symbol dy/dx for the derivative has the mental advantage of quickly reminding us of the whole process of forming the difference quotient $\Delta y/\Delta x$ and calculating its limit as $\Delta x \to 0$. It also has a practical advantage: certain fundamental formulas for computing derivatives are easier to remember and use when derivatives are written in Leibniz notation.

Yet this notation isn't perfect. For example, suppose that we want to write the numerical value of the derivative at a specific point, say $x = 3$. Because dy/dx doesn't display the variable x in the convenient way that $f'(x)$ does, we're forced to use such awkward notation as

$$\left(\frac{dy}{dx}\right)_{x=3} \quad \text{or} \quad \left.\frac{dy}{dx}\right|_{x=3}.$$

The concise symbol $f'(3)$ is superior to these expressions.

Prime and Leibniz notation are used interchangeably in this book. Each is good in its own way and both are used widely in scientific and mathematical literature.

Problem 2.9 Use the three-step rule (Problem 2.3) to show that if $f(x) = ax^2 + bx + c$, then $f'(x) = 2ax + b$.

Solution

STEP 1:

$$\begin{aligned} f(x+\Delta x) - f(x) &= [a(x+\Delta x)^2 + b(x+\Delta x) + c] - [ax^2 + bx + c] \\ &= [ax^2 + 2ax\Delta x + a(\Delta x)^2 + bx + b\Delta x + c] - [ax^2 + bx + c] \\ &= 2ax\Delta x + a(\Delta x)^2 + b\Delta x \\ &= \Delta x(2ax + a\Delta x + b). \end{aligned}$$

STEP 2:

$$\frac{f(x+\Delta x) - f(x)}{\Delta x} = 2ax + a\Delta x + b.$$

STEP 3:

$$f'(x) = \lim_{\Delta x \to 0} (2ax + a\Delta x + b) = 2ax + b.$$

Problem 2.10 Use the general rule in Problem 2.9 to compute the indicated derivatives.
(a) Find $g'(x)$ if $g(x) = 50 - 8x^2$.
(b) Find $F'(x)$ if $F(x) = 20 - 72x$.
(c) Find $H'(x)$ if $H(x) = 5 - 10x + 15x^2$.
(d) Find dx/dy if $x = 3y^2 + 7y - 6$.
(e) Find dv/dx if $v = 7x(500 - x)$.
(f) Find $f'(x)$ if $f(x) = 5x(x + 5)$.
(g) Find $g'(x)$ if $g(x) = 3 - (4x + 5)^2$.

Solution
(a) $g(x) = 50 - 8x^2 = -8x^2 + 50$, so $a = -8$ and $b = 0$. Hence $g'(x) = 2 \cdot (-8)x + 0 = -16x$.

(b) $F(x) = 20 - 72x = -72x + 20$, so $a = 0$ and $b = -72$. Hence $F'(x) = 2 \cdot 0 \cdot x - 72 = -72$.

(c) Here $a = 15$ and $b = -10$, so $H'(x) = 2 \cdot 15 \cdot x - 10 = 30x - 10$.

(d) Here $a = 3$ and $b = 7$, so $dx/dy = 2 \cdot 3 \cdot y + 7 = 6y + 7$.

(e) $v = 7x(500 - x) = 3500x - 7x^2$, so $a = -7$ and $b = 3500$. Hence $dv/dx = 2(-7)x + 3500 = -14x + 3500$.

(f) $f(x) = 5x(x + 5) = 5x^2 + 25x$, so $a = 5$ and $b = 25$. Hence $f'(x) = 2 \cdot 5 \cdot x + 25 = 10x + 25$.

(g) $g(x) = 3 - (4x + 5)^2 = -16x^2 - 40x - 22$, so $a = -16$ and $b = -40$. Hence $g'(x) = 2 \cdot -16 \cdot x - 40 = -32x - 40$.

Problem 2.11 Find all points on the curve $y = f(x)$ at which the tangent is horizontal.
(a) $y = 6 - x^2$
(b) $y = x^2 - 6x + 9$
(c) $y = x(20 - x)$

Solution (a) We must find all points where $f'(x) = 0$; the tangents to the curve at such points have slope 0 and hence are horizontal. By Problem 2.9, $f'(x) = -2x$, which equals zero only when $x = 0$. The tangent is horizontal at only (0, 6).

(b) By Problem 2.9, $f'(x) = 2x - 6$, which vanishes only when $x = 3$. The tangent is horizontal at only (3, 0).

(c) By Problem 2.9, $f'(x) = -2x + 20$, which equals 0 only when $x = 10$. The tangent is horizontal at only (10, 100).

Problem 2.12 Use the three-step rule (Problem 2.3) to calculate $f'(x)$ if $f(x)$ is equal to the given expression.

(a) $5x - x^3$

(b) $2x^3 - 3x^2 + 6x - 5$

(c) $x - \frac{1}{x}$

(d) $\frac{x}{x+1}$

(e) $\frac{1}{x^2}$

(f) $\frac{1}{x^2+1}$

(g) $\frac{2x}{x^2-1}$

(h) $\sqrt{x-1}$

Solution

(a) STEP 1:

$$\begin{aligned} f(x+\Delta x) - f(x) &= [5(x+\Delta x) - (x+\Delta x)^3] - [5x - x^3] \\ &= [5x + 5\Delta x - x^3 - 3x^2\Delta x - 3x(\Delta x)^2 - (\Delta x)^3] - [5x - x^3] \\ &= \Delta x(5 - 3x^2 - 3x\Delta x - (\Delta x)^2). \end{aligned}$$

STEP 2:

$$\frac{f(x+\Delta x) - f(x)}{\Delta x} = 5 - 3x^2 - 3x\Delta x - (\Delta x)^2.$$

STEP 3:

$$f'(x) = 5 - 3x^2.$$

(b) STEP 1:

$$\begin{aligned} f(x+\Delta x) - f(x) &= [2x^3 + 6x^2\Delta x + 6x(\Delta x)^2 + 2(\Delta x)^3 - 3x^2 - 6x\Delta x - \\ &\quad 3(\Delta x)^2 + 6x + 6\Delta x - 5] - [2x^3 - 3x^2 + 6x - 5] \\ &= \Delta x(6x^2 + 6x\Delta x + 2(\Delta x)^2 - 6x - 3\Delta x + 6). \end{aligned}$$

STEP 2:

$$\frac{f(x+\Delta x) - f(x)}{\Delta x} = 6x^2 + 6x\Delta x + 2(\Delta x)^2 - 6x - 3\Delta x + 6.$$

STEP 3:

$$f'(x) = 6x^2 - 6x + 6.$$

(c) STEP 1:

$$f(x+\Delta x)-f(x)=\left(x+\Delta x-\frac{1}{x+\Delta x}\right)-\left(x-\frac{1}{x}\right)$$
$$=\Delta x+\frac{1}{x}-\frac{1}{x+\Delta x}$$
$$=\Delta x+\frac{\Delta x}{x(x+\Delta x)}$$
$$=\Delta x\left(1+\frac{1}{x(x+\Delta x)}\right).$$

STEP 2:

$$\frac{f(x+\Delta x)-f(x)}{\Delta x}=1+\frac{1}{x(x+\Delta x)}.$$

STEP 3:

$$f'(x)=1+\frac{1}{x^2}.$$

(d) STEP 1:

$$f(x+\Delta x)-f(x)=\left[\frac{x+\Delta x}{x+\Delta x+1}\right]-\left[\frac{x}{x+1}\right]$$
$$=\frac{(x+\Delta x)(x+1)-x(x+\Delta x+1)}{(x+\Delta x+1)(x+1)}$$
$$=\Delta x\left(\frac{1}{(x+\Delta x+1)(x+1)}\right).$$

STEP 2:

$$\frac{f(x+\Delta x)-f(x)}{\Delta x}=\frac{1}{(x+\Delta x+1)(x+1)}.$$

STEP 3:

$$f'(x)=\frac{1}{(x+1)^2}.$$

(e) STEP 1:

$$f(x+\Delta x)-f(x)=\left[\frac{1}{(x+\Delta x)^2}\right]-\left[\frac{1}{x^2}\right]$$
$$=\frac{x^2-(x+\Delta x)^2}{x^2(x+\Delta x)^2}$$
$$=-\Delta x\left(\frac{2x+\Delta x}{x^2(x+\Delta x)^2}\right).$$

STEP 2:

$$\frac{f(x+\Delta x)-f(x)}{\Delta x}=-\frac{2x+\Delta x}{x^2(x+\Delta x)^2}.$$

STEP 3:

$$f'(x)=-\frac{2x}{x^2\cdot x^2}=-\frac{2}{x^3}.$$

(f) STEP 1:

$$\begin{aligned}f(x+\Delta x)-f(x)&=\frac{1}{(x+\Delta x)^2+1}-\frac{1}{x^2+1}\\&=\frac{x^2+1-((x+\Delta x)^2+1)}{((x+\Delta x)^2+1)(x^2+1)}\\&=-\Delta x\left(\frac{2x+\Delta x}{((x+\Delta x)^2+1)(x^2+1)}\right).\end{aligned}$$

STEP 2:

$$\frac{f(x+\Delta x)-f(x)}{\Delta x}=-\frac{2x+\Delta x}{((x+\Delta x)^2+1)(x^2+1)}.$$

STEP 3:

$$f'(x)=-\frac{2x}{(x^2+1)^2}.$$

(g) STEP 1:

$$\begin{aligned}f(x+\Delta x)-f(x)&=\frac{2(x+\Delta x)}{(x+\Delta x)^2-1}-\frac{2x}{x^2-1}\\&=\frac{2(x+\Delta x)(x^2-1)-2x((x+\Delta x)^2-1)}{((x+\Delta x)^2-1(x^2-1)}\\&=2\frac{(x^3+x^2\Delta x-x-\Delta x)-(x^3+2x^2\Delta x+x(\Delta x)^2-x)}{[(x+\Delta x)^2-1](x^2-1)}\\&=-2\Delta x\left(\frac{x^2+1+x\Delta x}{[(x+\Delta x)^2-1](x^2-1)}\right).\end{aligned}$$

STEP 2:

$$\frac{f(x+\Delta x)-f(x)}{\Delta x}=-2\left(\frac{x^2+1+x\Delta x}{[(x+\Delta x)^2-1](x^2-1)}\right).$$

STEP 3:

$$f'(x)=-\frac{2(x^2+1)}{(x^2-1)^2}.$$

(h) STEP 1: Using the algebraic trick in Problem 2.7,

$$
\begin{aligned}
f(x+\Delta x)-f(x) &= \sqrt{(x+\Delta x)-1}-\sqrt{x-1} \\
&= (\sqrt{(x+\Delta x)-1}-\sqrt{x-1})\cdot\frac{\sqrt{(x+\Delta x)-1}+\sqrt{x-1}}{\sqrt{(x+\Delta x)-1}+\sqrt{x-1}} \\
&= \frac{(x+\Delta x)-1-(x-1)}{\sqrt{(x+\Delta x)-1}+\sqrt{x-1}} \\
&= \Delta x\left(\frac{1}{\sqrt{(x+\Delta x)-1}-\sqrt{x-1}}\right).
\end{aligned}
$$

STEP 2:

$$\frac{f(x+\Delta x)-f(x)}{\Delta x}=\frac{1}{\sqrt{(x+\Delta x)-1}-\sqrt{x-1}}.$$

STEP 3:

$$f'(x)=\frac{1}{2\sqrt{x-1}}.$$

Problem 2.13 Consider the part of the curve $y = 1/x$ that lies in the first quadrant, and draw the tangent at an arbitrary point (x_0, y_0) on this curve.

(a) Show that the portion of the tangent which is cut off by the axes is bisected by the point of tangency.

(b) Find the area of the triangle formed by the axes and the tangent, and verify that this area is independent of the location of the point of tangency.

Solution (a) By Problem 2.5, the slope of the tangent is

$$-\frac{1}{x_0^2}=-y_0^2.$$

Hence the equation of the tangent is

$$\frac{y-y_0}{x-x_0}=-y_0^2;$$

that is,

$$y-y_0=-y_0^2x+x_0y_0^2.$$

Because

$$y_0=\frac{1}{x_0},$$

$x_0y_0^2 = y_0$ and the equation simplifies to $y = y_0(2 - xy_0)$.

The intercepts are $(0, 2y_0)$ and

$$\left(\frac{2}{y_0},0\right)=(2x_0,0).$$

Their midpoint is

$$\left(\frac{1}{2}(0+2x_0),\frac{1}{2}(2y_0+0)\right)=(x_0,y_0),$$

as required.

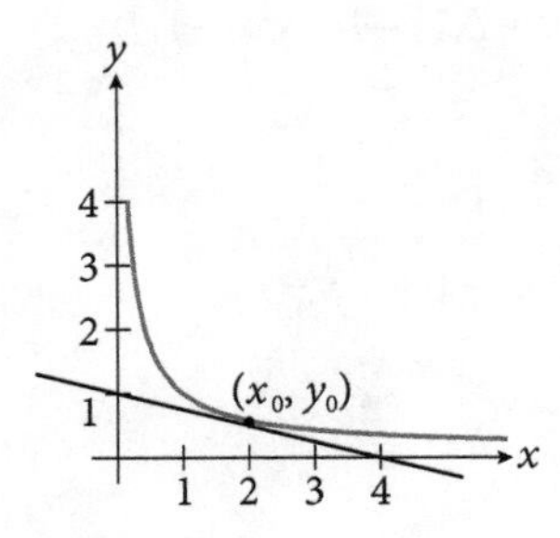

(b) The triangle has base $2x_0$ and height $2y_0$, so its area is

$$\frac{1}{2}\cdot 2x_0\cdot 2y_0=2x_0y_0=2x_0\frac{1}{x_0}=2.$$

Problem 2.14 Graph the function $y = f(x) = |x| + x$, and prove that this function is not differentiable at $x = 0$. Hint: In formula (1) in Problem 2.1, first take Δx positive, obtaining one limiting value; then take Δx negative, obtaining a different limiting value. In a situation of this kind we say that the function has a **right derivative** and a **left derivative**, but not a derivative.

Solution

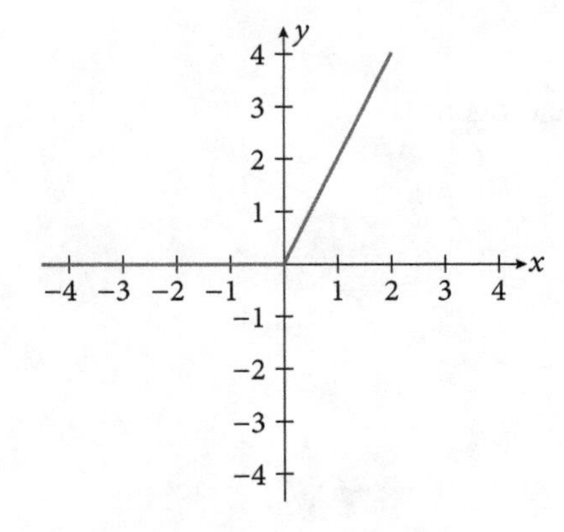

For $x = 0$ and $\Delta x > 0$,

$$\frac{f(x+\Delta x)-f(x)}{\Delta x}=\frac{f(\Delta x)-f(0)}{\Delta x}=\frac{2\Delta x-0}{\Delta x}=2.$$

For $x = 0$ and $\Delta x < 0$,

$$\frac{f(x+\Delta x)-f(x)}{\Delta x} = \frac{0-0}{\Delta x} = 0.$$

Hence

$$\lim_{\Delta x\to 0} \frac{f(x+\Delta x)-f(x)}{\Delta x}$$

does not exist.

Problem 2.15 If $g(x) = 3 - 2x^3$, then find $g'(0)$ and use it to compute the equation of the tangent line to the curve $y = 3 - 2x^3$ at the point (0, 3).

Solution When $x = 0$,

$$\frac{g(x+\Delta x)-g(x)}{\Delta x} = \frac{g(\Delta x)-g(0)}{\Delta x} = \frac{(3-2(\Delta x)^3)-3}{\Delta x} = -2(\Delta x)^2.$$

So

$$g'(0) = \lim_{\Delta x\to 0} \frac{g(0+\Delta x)-g(0)}{\Delta x} = \lim_{\Delta x\to 0}(-2(\Delta x)^2) = 0,$$

and the tangent line is horizontal. Because it passes through the point (0, 3), the equation for the tangent is $y = 3$.

Problem 2.16 Find the point on the graph of $y = x^2$ that is closest to the point (0, 3). Hint: Draw the graph, let (a, a^2) be the point, and find a as a root of a certain cubic equation that can be solved by inspection.

Solution This problem amounts to minimizing the distance between points (x, x^2) on the curve and (0, 3) or, equivalently, the square of this distance. In general, the square of the distance from (x, x^2) to (0, 3) is

$$\begin{aligned} f(x) &= \left(\sqrt{(x-0)^2+(x^2-3)^2}\right)^2 \\ &= x^2+(x^2-3)^2 \\ &= x^4-5x^2+9. \end{aligned}$$

From geometric considerations, a minimum distance exists. It must occur at a point on the curve $f(x) = x^4 - 5x^2 + 9$ whose tangent has zero slope, where $f'(x) = 0$. We compute $f'(x)$:

$$\begin{aligned} \frac{f(x+\Delta x)-f(x)}{\Delta x} &= \frac{(x+\Delta x)^4-5(x+\Delta x)^2+9-(x^4-5x^2+9)}{\Delta x} \\ &= 4x^3+6x^2\Delta x+4x\Delta x^2+\Delta x^3-10x-5\Delta x^2 \end{aligned}$$

and $f'(x) = 4x^3 - 10x$.

Now $4x^3 - 10x = 2x(2x^2 - 5) = 0$ when $x = 0$ or $x = \pm\sqrt{5/2}$. So the roots of $f'(x)$ are 0 and $\pm\sqrt{5/2}$. Because $f(0) = 0^2 + (0^2 - 3)^2 = 9$ while

$$f(\pm\sqrt{5/2}) = \frac{5}{2} + \left(\frac{5}{2} - 3\right)^2 = \frac{11}{4} < 9,$$

the points on the curve $y = x^2$ closest to (0, 3) are $(\sqrt{5/2}, 5/2)$ and $(-\sqrt{5/2}, 5/2)$; the minimal distance is $\sqrt{11/4} = \sqrt{11}/2$.

Problem 2.17 Show that the following function is not differentiable at $x = 0$:

$$f(x) = \begin{cases} x & \text{if } x \text{ is rational,} \\ 0 & \text{if } x \text{ is irrational.} \end{cases}$$

Solution

$$\frac{f(0+\Delta x) - f(0)}{\Delta x} = \frac{f(\Delta x)}{\Delta x} = \begin{cases} 1 & \text{if } \Delta x \text{ is rational,} \\ 0 & \text{if } \Delta x \text{ is irrational.} \end{cases}$$

Because there exist both rationals and irrationals arbitrarily close to 0,

$$\lim_{\Delta x \to 0} \frac{f(0+\Delta x) - f(0)}{\Delta x}$$

does not exist.

---

3 Velocity and Rates of Change

Problem 3.1 *Free fall* A freely falling object, say a rock, is dropped from the edge of a cliff 400 feet high. It is known from experiments that this rock falls $s = 16t^2$ feet in t seconds. Compute the number of seconds it takes for the rock to hit the ground after it starts falling.

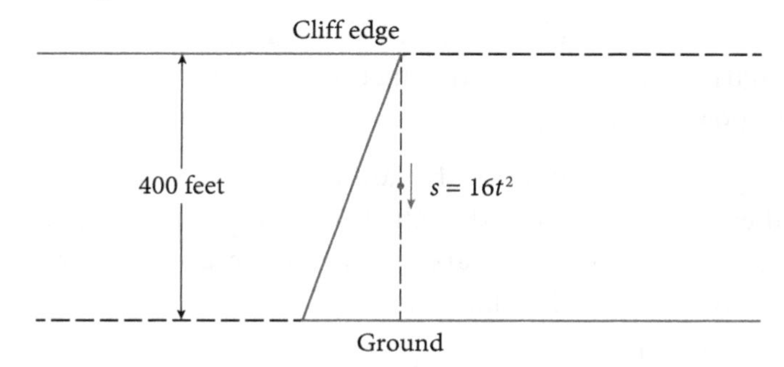

Solution The concept of the derivative (Chapter 2) is closely related to the problem of computing the velocity of a moving object. Consider a special case of the general velocity problem: that in which the object in question can be thought of as a point moving along a straight line, so that the position of the point is determined by a single coordinate s. The motion is fully known if we know where the moving point is at each moment; that is, if we know the position s as a function of the time t,

(1) $s = f(t)$.

The time is usually measured from some convenient initial moment $t = 0$.

0 P s s-axis

For this problem, the rock falls

(2) $s = 16t^2$

feet in t seconds, so when $t = 5$, $s = 400$. The rock therefore hits the ground 5 seconds after it starts to fall, and formula (2) is valid only for $0 \le t \le 5$.

Problem 3.2 *Average velocity* Compute the average velocity of a driver who travels a distance of 200 miles in 5 hours.

Solution Velocity, in its everyday sense, is a number measuring the rate at which distance is being traversed. We speak of walking 3 miles per hour (mi/h), driving 60 mi/h, and so on. We also speak of **average velocity**, and this is the number that we usually compute. If a driver travels a distance of 200 mi in 5 hours, then his average velocity is 40 mi/h, because

$$\frac{\text{distance traveled}}{\text{elapsed time}} = \frac{200 \text{ mi}}{5 \text{ h}} = 40 \text{ mi/h}.$$

In general,

$$\text{average velocity} = \frac{\text{distance traveled}}{\text{elapsed time}},$$

and this is a formula that almost everyone knows.

Problem 3.3 *Definition of velocity* As in Problem 3.1, a rock is dropped from the edge of a cliff 400 feet high. The position function for the falling rock is $f(t) = 16t^2$. Compute the rock's position, average velocity, and velocity after 1, 2, and 3 seconds of fall.

Solution The position function for the falling rock, $f(t) = 16t^2$, tells us that in the first second after the rock is released it falls $f(1) = 16$ ft, in the first 2 seconds $f(2) = 64$ ft, in the first 3 seconds $f(3) = 144$ ft, and so on. The average velocities during each of the first 3 seconds of fall are therefore

$$\frac{16}{1} = 16 \text{ ft/s}, \quad \frac{64-16}{1} = 48 \text{ ft/s}, \quad \text{and} \quad \frac{144-64}{1} = 80 \text{ ft/s}.$$

The rock is clearly falling faster and faster from moment to moment, but we still don't know exactly how fast it's falling at any given instant.

To find the velocity v of the rock at a given instant t, we proceed as follows. In the time interval of length Δt between t and a slightly later instant $t + \Delta t$, the rock falls a distance Δs.

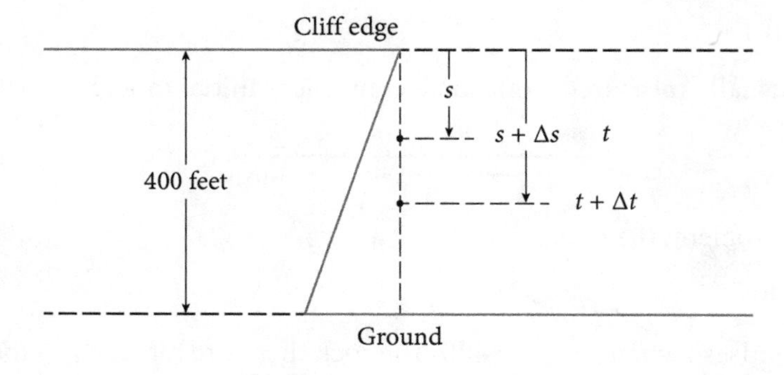

The average velocity during this interval is the quotient $\Delta s/\Delta t$. When Δt is small, this average velocity is close to the exact velocity v at the beginning of the interval; that is

$$v \cong \frac{\Delta s}{\Delta t},$$

where the symbol $\cong$ is read "is approximately equal to". Further, as Δt is made smaller and smaller, this approximation gets better and better, so we have

$$(3) \quad v = \lim_{\Delta t \to 0} \frac{\Delta s}{\Delta t}.$$

Our point of view here is that the velocity v is an intuitive concept, and (3) shows us how to compute it. However, it's also possible to regard (3) as the *definition* of the velocity, with the preceding remarks serving as motivation. The limit in (3) is clearly the derivative ds/dt, and carrying out the details we have

$$\begin{aligned} v = \frac{ds}{dt} &= \lim_{\Delta t \to 0} \frac{\Delta s}{\Delta t} \\ &= \lim_{\Delta t \to 0} \frac{16(t + \Delta t)^2 - 16t^2}{\Delta t} \\ &= \lim_{\Delta t \to 0} (32t + 16\Delta t) \\ &= 32t. \end{aligned}$$

This formula tells us that the velocity of the rock after 1, 2, and 3 seconds of fall is 32, 64, and 96 ft/s, and also that the rock hits the ground at 160 ft/s. We notice that the velocity increases by 32 ft/s during each second of fall. This fact is expressed by saying that the acceleration of the rock is 32 feet per second per second (ft/s^2), or, in the metric system, 9.80 meters per second per second (m/s^2).

The reasoning used in this example is valid for any motion along a straight line. For the general motion $s = f(t)$ [(1) in Problem 3.1], we therefore calculate the velocity v at time t in exactly the same way; that is, we approximate v more and more closely by the average velocity over a shorter and shorter interval of time beginning at the instant t:

$$v = \lim_{\Delta t \to 0} \frac{\Delta s}{\Delta t} = \lim_{\Delta t \to 0} \frac{f(t + \Delta t) - f(t)}{\Delta t}.$$

We recognize this as the derivative of the function $s = f(t)$, so the **velocity** of a point moving on a straight line is simply the derivative of its position function,

$$v = \frac{ds}{dt} = f'(t).$$

Sometimes this quantity is called the **instantaneous velocity** to emphasize that it's calculated at an instant t. However, after this point has been made, it's customary to omit the adjective.

Problem 3.4 *Velocity vs. speed* Describe the difference between velocity and speed.

Solution The terms "velocity" and "speed" are used interchangeably in everyday speech, but in mathematics and physics it's useful to distinguish them from one another. The **speed** of a point is defined to be the absolute value of the velocity,

$$\text{speed} = |v| = \left|\frac{ds}{dt}\right| = |f'(t)|.$$

The velocity can be positive or negative, depending on whether the point is moving along the line in the positive or negative direction; but the speed, being the magnitude of the velocity, is always positive or zero. The concept of speed is particularly useful in studies of motion along curved paths, for it tells us how fast the point is moving regardless of its direction. In our everyday experience, we learn the speed of a car at any moment by looking at the speedometer.

Problem 3.5 *Projectile path* Describe the complete path, including maximum height, of a projectile fired straight up from the ground with an initial velocity of 128 ft/s, where the height of the projectile during its flight is given by the formula $s = f(t) = 128t - 16t^2$.

Solution This projectile moves up and down along a straight line. However, the two parts of its path are shown slightly separated in the following figure, for the sake of visual clarity.

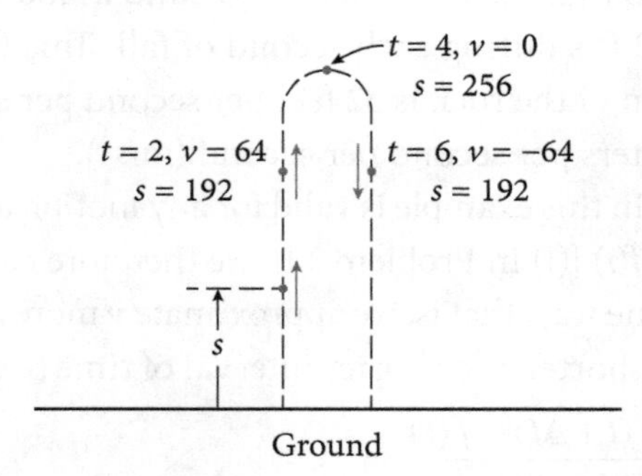

Let $s = f(t)$ be the height in feet of the projectile t seconds after firing. If the force of gravity were absent, then the projectile would continue moving upward with a constant velocity of 128 ft/s, and we would have $s = f(t) = 128t$. However, the action of gravity causes it to slow down, stop momentarily at the top of its flight, and then fall back to earth with increasing speed. Experimental evidence suggests that the height of the projectile during its flight is given by the formula

(4) $s = f(t) = 128t - 16t^2$.

If we write this equation in the factored form $s = 16t(8 - t)$, then we see that $s = 0$ when $t = 0$ and when $t = 8$. The projectile therefore returns to earth 8 seconds after it starts up, and (4) is valid only for $0 \le t \le 8$.

To learn more about the nature of this motion, it's necessary to know the velocity. If the general rule for computing derivatives of second-degree polynomials is applied to (4), then we find that the velocity at time t is

$$(5) \quad v = ds/dt = 128 - 32t.$$

At the top of its flight the projectile is momentarily at rest, and therefore $v = 0$. By (5), $t = 4$ when $v = 0$; and by (4), $s = 256$ when $t = 4$. In this way we find the maximum height reached by the projectile and the time required to reach this height (see the preceding figure). As t increases from 0 to 8, it's clear from (5) that v decreases from 128 ft/s to −128 ft/s; in fact, v decreases by 32 ft/s during each second of flight, and this is expressed by saying that the acceleration is −32 feet per second per second (ft/s^2). Note that the velocity is positive from $t = 0$ to $t = 4$, when s is increasing; and it is negative from $t = 4$ to $t = 8$, when s is decreasing. In particular, it's easy to see from (5) that $v = 64$ ft/s when $t = 2$ and $v = -64$ ft/s when $t = 6$ (the speed is 64 ft/s at both times).

Problem 3.6 *Rate of change* Let $y = f(x)$. Give the formula for the rate of change of y with respect to x.

Solution Velocity is an example of the concept of rate of change, which is basic for all the sciences. For any function $y = f(x)$, the derivative dy/dx is called the **rate of change** of y with respect to x. Intuitively, this is the change in y that would be produced by an increase of one unit in x if the rate of change remained constant.

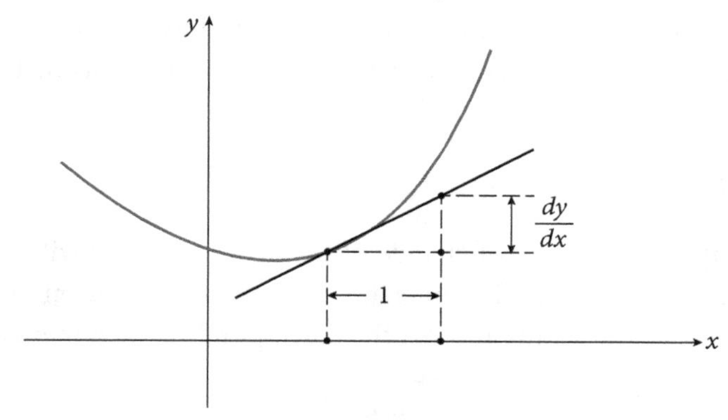

In this terminology, velocity is simply the rate of change of position with respect to time. When time is the independent variable, we often omit the phrase "with respect to time" and speak only of the "rate of change".

Problem 3.7 *Examples of rates of change* Describe the rate of change of

(a) the velocity v of an object with respect to time t;

(b) the volume V and the depth x of water being pumped into the conical tank at the rate of 5 ft^3/min, with respect to time t;

(c) the area A of a circle with respect to its radius r.

Solution (a) We know that velocity is important in studying the motion of a point along a straight line, but the way the velocity changes is also important. By definition, the **acceleration** of a moving point is the rate of change of its velocity v,

$$a = \frac{dv}{dt}.$$

(b) Water is being pumped into the conical tank at the rate of 5 ft^3/min.

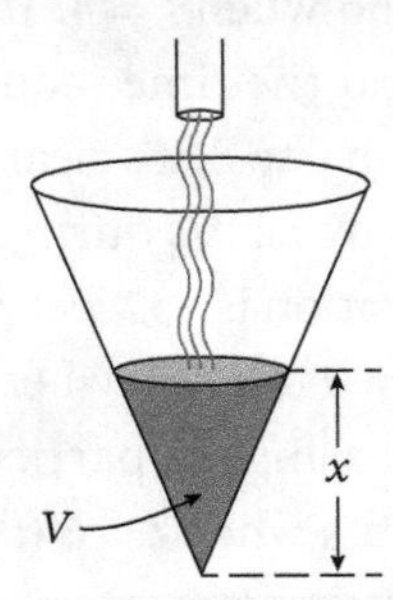

If V denotes the volume of water in the tank at time t, then

$$\frac{dV}{dt} = 5.$$

The rate of change of the depth x is the derivative dx/dt, and this is not constant. It's intuitively clear that this rate of change is large when the area of the surface of the water is small, and becomes smaller as this area increases.

(c) We know that the area A of a circle in terms of its radius r is given by the formula $A = \pi r^2$, and the derivative of this function is easy to compute by using the three-step rule (Problem 2.3):

$$\frac{dA}{dr} = 2\pi r. \tag{6}$$

This equation says that the rate of change of the area of a circle with respect to its radius equals its circumference. To understand the geometric reason for this fact, let Δr be an increment of the radius and ΔA the corresponding increment of the area.

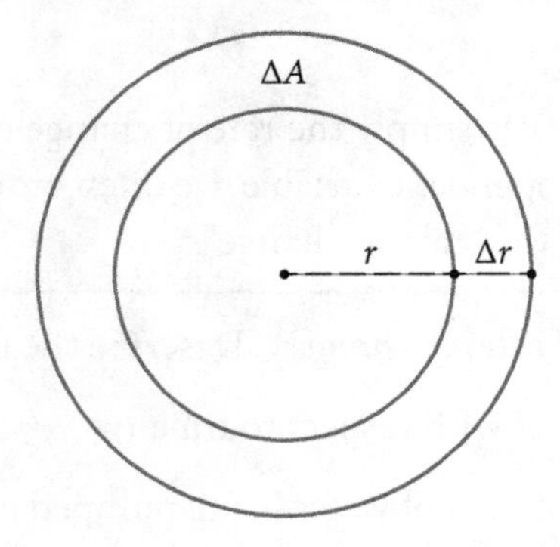

It's clear that ΔA is the area of the narrow band around the circle, and this is approximately the product of the circumference $2\pi r$ and the width Δr of the band. The difference quotient $\Delta A/\Delta r$ is therefore close to $2\pi r$, and by letting $\Delta r \to 0$ we obtain (6) by geometric reasoning.

Problem 3.8 *Marginal cost* In economics, **marginal cost** is the change in the total cost that arises when the quantity produced is incremented by one unit; that is, it's the cost of producing one more unit of a product. Describe this concept as a rate of change.

Solution The rate of change of a quantity Q with respect to a suitable independent variable is called **marginal** Q. Thus we have marginal cost, marginal revenue, marginal profit, and so on. If $C(x)$ is the cost of manufacturing x units of a product, then the marginal cost is dC/dx. In most cases x is a large number, so 1 is small compared with x and dC/dx is approximately $C(x + 1) - C(x)$.

Problem 3.9 According to Problem 2.9, the general quadratic function

$$s = f(t) = at^2 + bt + c$$

has derivative

$$ds/dt = f'(t) = 2at + b.$$

Each of the formulas below describes the motion of a point along a horizontal line whose positive direction is to the right. In each case, (i) use the formula just stated to compute the velocity $v = ds/dt$ by inspection; (ii) find the times when the velocity is zero, so that the point is momentarily at rest; (iii) find the times when the point is moving to the right.

(a) $s = 3t^2 - 12t + 7$
(b) $s = 2t^2 + 28t - 6$
(c) $s = 7t^2 + 2$
(d) $s = (2t - 6)^2$

Solution (a) (i) $v = ds/dt = (3t^2 - 12t + 7)' = 6t - 12$. (ii) $v = 0$ when $6t - 12 = 0$; that is, when $t = 2$. (iii) $v > 0$ when $t > 2$.

(b) (i) $v = ds/dt = (2t^2 + 28t - 6)' = 4t + 28$. (ii) $v = 0$ when $4t + 28 = 0$; that is, when $t = -7$. (iii) $v > 0$ when $t > -7$.

(c) (i) $v = ds/dt = (7t^2 + 2)' = 14t$. (ii) $v = 0$ when $14t = 0$; that is, when $t = 0$. (iii) $v > 0$ when $t > 0$.

(d) (i) $v = ds/dt = (4t^2 - 24t + 36)' = 8t - 24$. (ii) $v = 0$ when $8t - 24 = 0$; that is, when $t = 3$. (iii) $v > 0$ when $t > 3$.

Problem 3.10 A coin falls off a ledge of a building and falls to the ground below. The ledge is 784 ft above the ground. The coin falls a distance of $s = 16t^2$ feet in t seconds.
(a) How long does the coin fall before it hits the ground?
(b) What is the average velocity at which the coin falls during the first 3 seconds?
(c) What is the average velocity at which the coin falls during the last 3 seconds?
(d) What is the instantaneous velocity of the coin when it hits the ground?

Solution (a) The coin falls 784 ft. So at the time of impact $s = 748 = 16t^2$, $t^2 = 49$, and $t = \pm 7$ sec. The coin falls for 7 seconds.

(b) During the first 3 seconds, the coin falls $s = 16 \cdot 3^2 = 144$ ft, so the average velocity during this time is $144/3 = 48$ ft/s.

(c) By part (a), the last 3 seconds correspond to the time interval from $t = 4$ to $t = 7$ seconds. At $t = 4$, $s = 16 \cdot 4^2 = 256$ ft, and we know $s = 784$ ft at $t = 7$. So the average velocity during this interval is $(784 - 256)/3 = 176$ ft/s.

(d) By Problem 2.9, the instantaneous velocity of the coin is $v = ds/dt = (16t^2)' = 32t$, valid for $0 \le t \le 7$. Hence when $t = 7$, $v = 32 \cdot 7 = 224$ ft/s.

Problem 3.11 Consider the function $y = 3x^2 + 4$.

(a) Find the average rate of change of y with respect to x between the points $x = 1$ and $x = 3$.

(b) Find the instantaneous rate of change of y with respect to x at the point $x = 1$.

(c) Find the instantaneous rate of change of y with respect to x at the point $x = 3$.

Solution (a) When $x = 1$, $y = 3 \cdot 1^2 + 4 = 7$ and when $x = 3$, $y = 3 \cdot 3^2 + 4 = 31$. So $\Delta y = 31 - 7 = 24$ and $\Delta x = 3 - 1 = 2$. The average rate of change of y with respect to x between $x = 1$ and $x = 3$ is thus $\Delta y/\Delta x = 24/2 = 12$.

(b) The instantaneous rate of change of y with respect to x is the derivative $y = (3x^2 + 4)' = 6x$, using Problem 2.9. When $x = 1$, this rate is $6 \cdot 1 = 6$.

(c) By part (b), the rate is $6 \cdot 3 = 18$ when $x = 3$.

Problem 3.12 Starting from rest, a car moves s feet in t seconds where $s = 4.4t^2$. How long does it take the car to reach the velocity of 60 mi/h (= 88 ft/s)?

Solution $v = (4.4t^2)' = 2(4.4t) = 8.8t$. Hence $v = 88$ at $t = 10$. It takes 10 seconds.

Problem 3.13 An oil tank is to be drained for cleaning. If there are V gallons of oil left in the tank t minutes after the draining begins, where $V = 40(50 - t)^2$, then find

(a) the average rate at which oil drains out of the tank during the first 20 minutes;

(b) the rate at which oil is flowing out of the tank 20 minutes after the draining begins.

Solution (a) $V = 40(50 - t)^2$. At $t = 0$, $V = 40 \cdot 50^2 = 100000$. At $t = 20$, $V = 40 \cdot 30^2 = 36000$. The tank therefore loses $100000 - 36000 = 64000$ gallons in the first 20 minutes, so the average rate is $64000/2 = 3200$ gal/min.

(b) $v = 100000 - 4000t + 40t^2$, so $dV/dt = -4000 + 80t$. The rate at which oil is flowing out is $-dV/dt = -4000 - 80t$. At $t = 20$, this rate is $4000 - 80 \cdot 20 = 2400$ gal/min.

Problem 3.14 Suppose that a balloon of volume V and radius r is being inflated, so that V and r are both functions of the time t. If dV/dt is constant, what can be said (without calculation) about the behavior of dr/dt as r increases?

Solution If the radius of a balloon is large, then it takes a smaller increase in radius to allow a given volume increase than if the radius is small. Because the rate of volume growth is fixed, we see that dr/dt decreases as r increases.

Problem 3.15 A coin is thrown straight up from the roof of a 200-foot building. After t seconds, it is

$$s = 200 + 24t - 16t^2$$

feet above the ground. When does the coin start to fall? What is its speed when it has fallen 1 foot?

Solution The coin's velocity is $v = ds/dt = (24 - 32t)$ ft/s. It starts to fall when $v = 0$; that is, when $t = 3/4$. When it has fallen 1 ft, we have $s(t) = s(3/4) - 1$; that is, $200 + 24t - 16t^2 = (200 + 24(3/4) - 16(3/4)^2) - 1$. This equation simplifies to $2t^2 - 3t + 1 = 0$; that is, $t = 1/2$ or $t = 1$. At $t = 1/2$, the coin hasn't started to fall, so we have $t = 1$. The speed then is $|v(t)| = |24 - 32(1)| = 8$ ft/s.

Problem 3.16 Use the three-step rule (Problem 2.3) to show that the rate of change of the volume V of a sphere with respect to its radius r equals the surface area A. (The volume of a sphere is $(4/3)\pi r^3$ and the surface area is $4\pi r^2$.)

Solution

STEP 1:

$$V(r+\Delta r)-V(r)=\frac{4}{3}\pi(r+\Delta r)^3-\frac{4}{3}\pi r^3=\frac{4}{3}\pi\Delta r(3r^2+3r\Delta r+(\Delta r)^2)$$

STEP 2:

$$\frac{V(r+\Delta r)-V(r)}{\Delta r}=\frac{4}{3}\pi(3r^2+3r\Delta r+(\Delta r)^2)$$

STEP 3:

$$V'(r)=\lim_{\Delta r\to 0}\frac{4}{3}\pi(3r^2+3r\Delta r+(\Delta r)^2)=\frac{4}{3}\pi(3r^2)=4\pi r^2=A$$

---

4 Limits

Problem 4.1 *Definition of the limit* The definition of the derivative (Problem 2.1) relies on the concept of the limit of a function (which was used with little explanation in the preceding chapters). Define the limit $\lim_{x\to a} f(x) = L$ precisely.

Solution Consider a function $f(x)$ that is defined for all values of x near a point a on the x-axis but not necessarily at a itself. Suppose that there exists a real number L with the property that $f(x)$ gets closer and closer to L as x gets closer and closer to a.

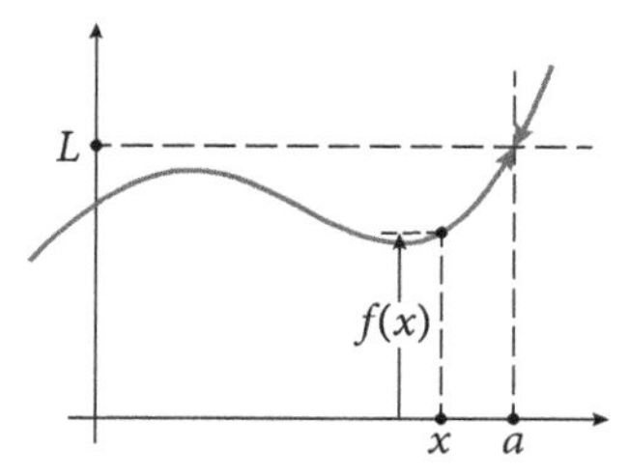

Under these circumstances we say that L is the **limit** of $f(x)$ as x approaches a, and we denote this symbolically by

(1) $\lim\limits_{x\to a} f(x) = L.$

If no real number L with this property exists, then we say that $f(x)$ **has no limit** as x approaches a, or that $\lim_{x\to a} f(x)$ **does not exist**. Another notation equivalent to (1) is

$$f(x) \to L \quad \text{as} \quad x \to a,$$

which is read "$f(x)$ approaches L as x approaches a". In thinking about the meaning of (1), it's essential to understand that it doesn't matter what happens to $f(x)$ when x *equals* a; all that matters is the behavior of $f(x)$ for x's that are *near* a.

These informal descriptions of the meaning of (1) are helpful intuitively and are adequate for most practical purposes. Nevertheless, they're too imprecise to be acceptable as definitions, owing to the vagueness of such expressions as "closer and closer" and "approaches". The exact meaning of (1) is too important to be left mainly to the imagination, and a rigorous, precise definition is necessary.

We begin by analyzing a specific example with the goal of determining the general situation:

$$\lim_{x\to 0}\frac{2x^2+x}{x}=1.$$

Here the function that we must examine is

$$y=f(x)=\frac{2x^2+x}{x}.$$

This function is not defined for $x = 0$, and for $x \neq 0$ its values are given by the simpler expression

$$f(x)=\frac{x(2x+1)}{x}=2x+1.$$

If we examine the graph, then it's clear that $f(x)$ is close to 1 when x is close to 0.

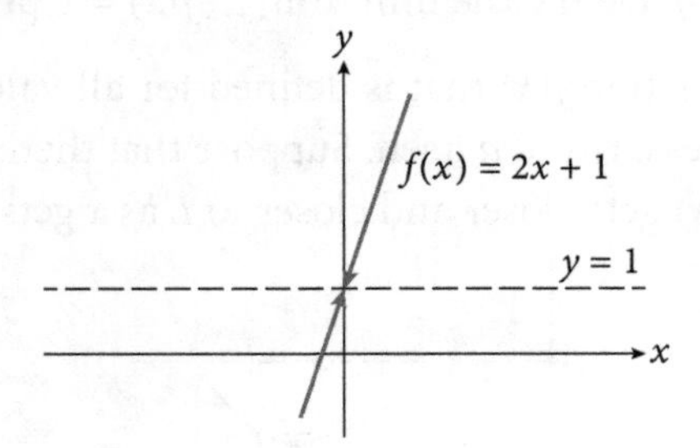

To give a quantitative description of this qualitative behavior, we need a formula for the difference between $f(x)$ and the limiting value 1:

$$f(x)-1=(2x+1)-1=2x.$$

We see from this formula that $f(x)$ can be made *as close as we like* to 1; that is, this difference can be made *as small as we like*, by taking x *sufficiently close* to 0. Thus,

$$f(x)-1=\tfrac{1}{100}\quad\text{when}\quad x=\tfrac{1}{200},$$
$$f(x)-1=\tfrac{1}{1000}\quad\text{when}\quad x=\tfrac{1}{2000},$$

and so on. More generally, let ε (epsilon) be any positive number given in advance, no matter how small, and define δ (delta) by $\delta = \varepsilon/2$. Then the distance from $f(x)$ to 1 will be smaller than ε, provided only that the distance from x to 0 is smaller than δ; that is,

$$\text{if}\quad |x|<\delta=\tfrac{1}{2}\varepsilon,\quad\text{then}\quad |f(x)-1|=2|x|<\varepsilon.$$

This assertion is much more precise than the vague statement that $f(x)$ is "close" to 1 when x is "close" to 0. It tells us exactly how close x must be to 0 to guarantee that $f(x)$ will attain a previously specified degree of closeness to 1. Of course, x isn't permitted to equal 0 here, because $f(x)$ has no meaning for $x = 0$.

The **epsilon-delta definition** of the meaning of (1) should now be easy to grasp. The defining condition is:

For each positive number ε there exists a positive number δ with the property that

$$|f(x) - L| < \varepsilon$$

for any number $x \neq a$ that satisfies the inequality

$$|x - a| < \delta.$$

In words: if $\varepsilon > 0$ is given, then $\delta > 0$ can be found with the property that $f(x)$ will be "ε-close" to L whenever x is "δ-close" to a. As usual, we're concerned with the behavior of $f(x)$ only *near* the point $x = a$, and not at all with what happens *at* $x = a$.

It's helpful to interpret these ideas in terms of the graph of the function $y = f(x)$, as shown in the following figure.

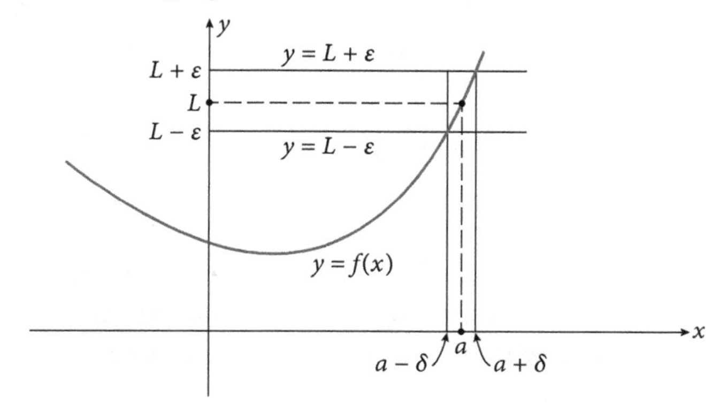

In this figure, 2ε is the width of the horizontal strip centered on the line $y = L$, 2δ is the width of the vertical strip centered on the line $x = a$, and the defining condition stated above can be expressed this way:

For each horizontal strip, no matter how narrow, there exists a vertical strip such that if $x \neq a$ is confined to the vertical strip, then the corresponding part of the graph will be confined to the horizontal strip.

The precise definition of the meaning of (1) plays a crucial role in the theory of calculus, but an intuitive understanding of limits is more than adequate for most purposes.

Problem 4.2 Evaluate the following limits.

(a) $\lim_{x\to 2}(3x + 4)$

(b) $\lim\limits_{x\to 1}\dfrac{x^2 - 1}{x - 1}$

Solution (a) It's clear that as x approaches 2, $3x$ approaches 6, and $3x + 4$ approaches $6 + 4 = 10$, so

$$\lim_{x\to 2}(3x + 4) = 10.$$

(b) First notice that the function $(x^2 - 1)/(x - 1)$ is undefined at $x = 1$, because both numerator and denominator equal 0. But this fact is irrelevant because all that matters is the behavior of the function for x's that are *near* 1 but different from 1, and for all such x's the following cancellation is valid, the function equals $x + 1$, and this is near 2.

$$\lim_{x\to 1}\frac{x^2-1}{x-1} = \lim_{x\to 1}\frac{(x+1)(x-1)}{x-1} = \lim_{x\to 1}(x+1) = 2.$$

Problem 4.3 Evaluate the limits

$$\lim_{x\to 0}\frac{x}{|x|}, \quad \lim_{x\to 0}\frac{1}{x}, \quad \text{and} \quad \lim_{x\to 0}\frac{1}{x^2}.$$

Solution The behavior of these limits is most easily understood by inspecting the graphs of the functions $x/|x|$, $1/x$, and $1/x^2$.

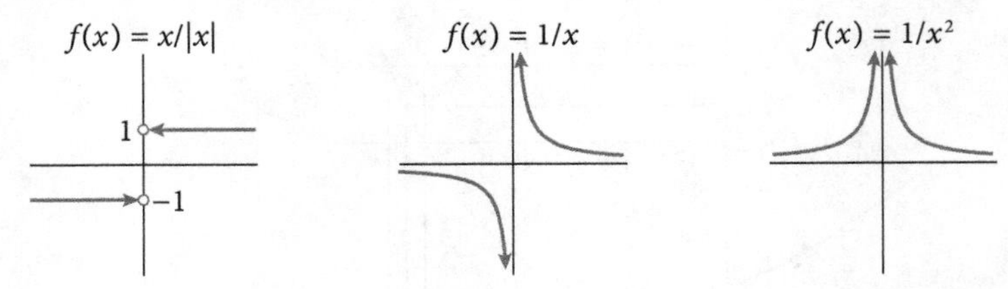

In the first case the function equals 1 when x is positive and −1 when x is negative (and is undefined for $x = 0$), so there is no single number that the values of the function approach as x approaches 0 without regard to sign. We can be more specific about the way that this limit fails to exist by writing

$$\lim_{x\to 0+}\frac{x}{|x|} = 1 \quad \text{and} \quad \lim_{x\to 0-}\frac{x}{|x|} = -1.$$

The notations $x \to 0+$ and $x \to 0-$ indicate that the variable x approaches 0 from the positive side (the right) and from the negative side (the left), respectively. The other two limits fail to exist because in each case the values of the function become arbitrarily large in absolute value as x approaches 0. In symbols,

$$\lim_{x\to 0+}\frac{1}{x} = \infty, \quad \lim_{x\to 0-}\frac{1}{x} = -\infty, \quad \text{and} \quad \lim_{x\to 0}\frac{1}{x^2} = \infty.$$

Remember that the number L in (1) in Problem 4.1 must be a *real* number; $L = \infty$ does not qualify.

Problem 4.4 *Limit laws* The main rules for calculating with limits, called **limit laws** or **limit theorems**, are exactly what we would expect. For example, it can be shown that

$$\lim_{x\to a} x = a$$

and, if c is a constant, then

$$\lim_{x\to a} c = c.$$

Write a formula that states that the limit of a sum is the sum of the limits, and similarly for differences, products, and quotients.

Solution If $\lim_{x\to a} f(x) = L$ and $\lim_{x\to a} g(x) = M$, then

$$\lim_{x\to a}[f(x)+g(x)] = L+M,$$

$$\lim_{x\to a}[f(x)-g(x)] = L-M,$$

$$\lim_{x\to a} f(x)g(x) = LM,$$

and

$$\lim_{x\to a}\frac{f(x)}{g(x)} = \frac{L}{M} \quad (\text{if } M \neq 0).$$

Problem 4.5 *Two trigonometric limits* Show that

$$(2)\quad \lim_{\theta\to 0}\frac{\sin\theta}{\theta} = 1$$

and

$$(3)\quad \lim_{\theta\to 0}\frac{1-\cos\theta}{\theta} = 0,$$

where θ is measured in radians.

Solution These two trigonometric limits are of crucial importance in calculus.

We can't simply set $\theta = 0$ in (2), because the result is the meaningless quotient 0/0. Note that this differs from an algebraic limit like

$$\lim_{x\to 0}\frac{3x^2+2x}{x} = \lim_{x\to 0}\frac{x(3x+2)}{x} = \lim_{x\to 0}(3x+2) = 2,$$

because there's no apparent way to cancel θ from $\sin\theta$. To assess what's happening in (2), let's calculate the numerical value of the ratio for several small values of θ. We begin by observing that if we replace θ by $-\theta$ in the ratio, then we have

$$\frac{\sin(-\theta)}{-\theta} = \frac{-\sin\theta}{-\theta} = \frac{\sin\theta}{\theta},$$

so we can restrict our attention to positive θ's. By using a calculator set to the radians mode, we can construct the following table of values correct to eight decimal places.

θ	$(\sin\theta)/\theta$	θ	$(\sin\theta)/\theta$
1	0.84147098	0.1	0.99833417
0.5	0.95885108	0.05	0.99958339
0.4	0.97354586	0.01	0.99998333
0.3	0.98506736	0.005	0.99999583
0.2	0.99334665	0.001	0.99999983

This numerical evidence strongly suggests (but doesn't prove) that

$$\lim_{\theta\to 0}\frac{\sin\theta}{\theta}=1.$$

We now establish (2) by a simple geometric argument. Let P and Q be two nearby points on a unit circle, and let $\overline{PQ}$ and $\overset{\frown}{PQ}$ denote the lengths of the chord and the arc connecting these points.

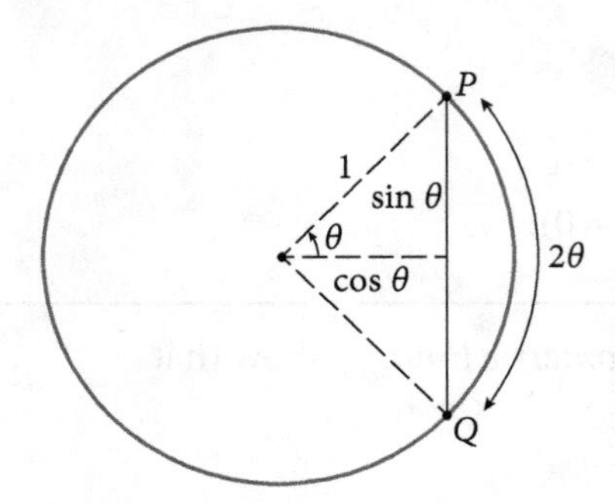

Then the ratio of the chord length to the arc length evidently approaches 1 as the two points move together:

$$\frac{\text{chord length } \overline{PQ}}{\text{arc length } \overset{\frown}{PQ}}\to 1 \quad \text{as} \quad \overset{\frown}{PQ}\to 0.$$

By using the notation in the figure, this geometric statement is equivalent to

$$\frac{2\sin\theta}{2\theta}=\frac{\sin\theta}{\theta}\to 1 \quad \text{as} \quad 2\theta\to 0 \quad \text{or} \quad \theta\to 0,$$

and this is (2).

Now we establish (3):

$$\lim_{\theta\to 0}\frac{1-\cos\theta}{\theta}=0.$$

This limit follows from (2) by the use of the trigonometric identity $\sin^2\theta+\cos^2\theta=1$:

$$\begin{aligned}
\lim_{\theta\to 0}\frac{1-\cos\theta}{\theta}&=\lim_{\theta\to 0}\left(\frac{1-\cos\theta}{\theta}\cdot\frac{1+\cos\theta}{1+\cos\theta}\right)\\
&=\lim_{\theta\to 0}\frac{1-\cos^2\theta}{\theta(1+\cos\theta)}\\
&=\lim_{\theta\to 0}\frac{\sin^2\theta}{\theta(1+\cos\theta)}\\
&=\lim_{\theta\to 0}\left(\frac{\sin\theta}{\theta}\cdot\frac{\sin\theta}{1+\cos\theta}\right)\\
&=\left(\lim_{\theta\to 0}\frac{\sin\theta}{\theta}\right)\left(\lim_{\theta\to 0}\frac{\sin\theta}{1+\cos\theta}\right)=1\cdot\frac{0}{1+1}=0.
\end{aligned}$$

The end of this calculation uses the facts that $\sin\theta \to 0$ and $\cos\theta \to 1$ as $\theta \to 0$, which are easily verified by examining the geometric meaning of $\sin\theta$ and $\cos\theta$ in the unit circle in the figure above.

The validity of statements (2) and (3) can be understood directly from geometry by thinking of θ as a small angle and inspecting the unit circle in the following figure.

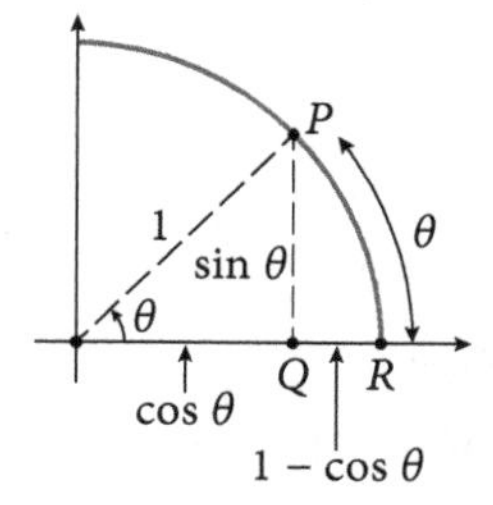

The definitions of sine and cosine tell us that $PQ = \sin\theta$, $PR = \theta$, and $QR = 1 - \cos\theta$. It's easy to see that the ratio $(\sin\theta)/\theta = PQ/PR$ is < 1 and close to 1, and visibly approaches 1 as θ approaches 0. The same type of inspection can also be applied to (3). This time the figure tells us that the ratio $(1 - \cos\theta)/\theta = QR/PR$ is a small number that clearly approaches 0 as θ approaches 0.

Problem 4.6 Evaluate the following limits.

(a) $\lim\limits_{x\to 3}(7x-6)$

(b) $\lim\limits_{x\to 0}\dfrac{5}{x-1}$

(c) $\lim\limits_{x\to 3}\dfrac{3x-9}{x-3}$

(d) $\lim\limits_{x\to 5}\dfrac{x-3-2x^2}{1+3x}$

(e) $\lim\limits_{x\to -3}\left(\dfrac{4x}{x+3}+\dfrac{12}{x+3}\right)$

(f) $\lim\limits_{x\to 7}\dfrac{x^2+x-56}{x^2-11x+28}$

(g) $\lim\limits_{x\to 0}\dfrac{x^2}{|x|}$

(h) $\lim\limits_{x\to 4}\dfrac{x-4}{x-\sqrt{x}-2}$

Solution

(a) $\lim\limits_{x\to 3}(7x-6) = \lim\limits_{x\to 3}7x - \lim\limits_{x\to 3}6 = \lim\limits_{x\to 3}(7)\lim\limits_{x\to 3}(x) - \lim\limits_{x\to 3}(6) = 7\cdot 3 - 6 = 15.$

(b) $\lim_{x\to 0}\frac{5}{x-1}=\frac{5}{\lim_{x\to 0}(x-1)}=\frac{5}{0-1}=-5.$

(c) $\lim_{x\to 3}\frac{3x-9}{x-3}=\lim_{x\to 3}3=3.$

(d) $\lim_{x\to 5}\frac{x-3-2x^2}{1+3x}=\frac{5-3-2\cdot 5^2}{1+3\cdot 5}=-3.$

(e) $\lim_{x\to -3}\left(\frac{4x}{x+3}+\frac{12}{x+3}\right)=\lim_{x\to -3}\left(\frac{4(x+3)}{x+3}\right)=\lim_{x\to -3}4=4.$

(f) $\lim_{x\to 7}\frac{x^2+x-56}{x^2-11x+28}=\lim_{x\to 7}\frac{(x-7)(x+8)}{(x-7)(x-4)}=\lim_{x\to 7}\frac{x+8}{x-4}=\frac{7+8}{7-4}=5.$

(g) $\lim_{x\to 0}\frac{x^2}{|x|}=\lim_{x\to 0}\frac{|x|^2}{|x|}=\lim_{x\to 0}|x|=0.$

(h) $\lim_{x\to 4}\frac{x-4}{x-\sqrt{x}-2}=\lim_{x\to 4}\frac{(\sqrt{x}-2)(\sqrt{x}+2)}{(\sqrt{x}-2)(\sqrt{x}+1)}=\lim_{x\to 4}\frac{\sqrt{x}+2}{\sqrt{x}+1}=\frac{\sqrt{4}+2}{\sqrt{4}+1}=\frac{4}{3}.$

Problem 4.7 If $\lim_{x\to a}f(x)=4$, $\lim_{x\to a}g(x)=-2$, and $\lim_{x\to a}h(x)=0$, then evaluate the following limits.

(a) $\lim_{x\to a}[f(x)-g(x)]$

(b) $\lim_{x\to a}[g(x)]^2$

(c) $\lim_{x\to a}\frac{f(x)}{g(x)}$

(d) $\lim_{x\to a}\frac{h(x)}{f(x)}$

(e) $\lim_{x\to a}\frac{f(x)}{h(x)}$

(f) $\lim_{x\to a}\frac{1}{[f(x)+g(x)]^2}$

Solution

(a) $\lim_{x\to a}[f(x)-g(x)]=4-(-2)=6.$

(b) $\lim_{x\to a}[g(x)]^2=(-2)^2=4.$

(c) $\lim_{x\to a}\frac{f(x)}{g(x)}=\frac{4}{-2}=-2.$

(d) $\lim_{x\to a}\frac{h(x)}{g(x)}=\frac{0}{4}=0.$

(e) Suppose that $\lim_{x \to a} \dfrac{f(x)}{h(x)}$ exists and equals L. Then, by the limit laws (Problem 4.4),

$$\lim_{x\to a} f(x) = \lim_{x\to a}\left(\frac{f(x)}{h(x)} h(x)\right) = \lim_{x\to a}\frac{f(x)}{h(x)} \cdot \lim_{x\to a} h(x) = L \cdot 0 = 0;$$

but this contradicts the fact that $\lim_{x \to a} f(x) = 4$. Hence $\lim_{x\to a}\dfrac{f(x)}{h(x)}$ does not exist.

(f) $\lim_{x\to a} \dfrac{1}{[f(x)+g(x)]^2} = \dfrac{1}{[4+(-2)]^2} = \dfrac{1}{4}$.

Problem 4.8 In many situations we're interested in the behavior of $f(x)$ when x is large and positive. If there exists a real number L with the property that $f(x)$ gets closer and closer to L as x gets larger and larger, then we say that L is the limit of $f(x)$ as x approaches infinity, and we symbolize this by writing $\lim_{x \to \infty} f(x) = L$.

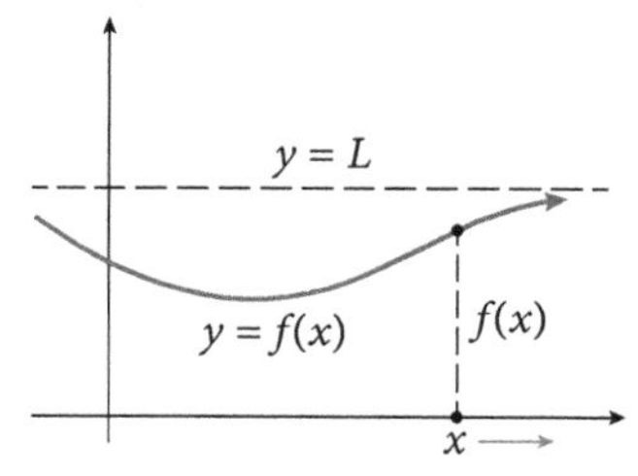

Evaluate the following limits.

(a) $\lim_{x\to\infty} \dfrac{1}{x}$

(b) $\lim_{x\to\infty} \dfrac{3x-2}{2x+1}$

Hint: Divide both numerator and denominator of this quotient by x.

(c) $\lim_{x\to\infty} \dfrac{3x^2-x-2}{5x^2+4x+1}$

(d) $\lim_{x\to\infty} (\sqrt{x^2+1}-x)$

(e) $\lim_{x\to\infty} \dfrac{\sqrt{x}+x^2}{2x-x^2}$

(f) $\lim_{x\to\infty} \dfrac{x^4-3x^2+x}{x^3-x+2}$

Solution

(a) Observe that when x is large, $1/x$ is small. In fact, by taking x large enough, we can make $1/x$ as close to 0 as we like. Therefore, we have $\lim_{x \to \infty} (1/x) = 0$.

(b)

$$\lim_{x\to\infty}\frac{3x-2}{2x+1}=\lim_{x\to\infty}\frac{(3x-2)/x}{(2x+1)/x}=\lim_{x\to\infty}\frac{3-\dfrac{2}{x}}{2+\dfrac{1}{x}}=\frac{\lim\limits_{x\to\infty}3-2\lim\limits_{x\to\infty}\dfrac{1}{x}}{\lim\limits_{x\to\infty}2+\lim\limits_{x\to\infty}\dfrac{1}{x}}=\frac{3-2(0)}{2+0}=\frac{3}{2}.$$

(c) As x becomes large, both numerator and denominator become large, so it's not obvious what happens to their ratio. We must do some preliminary algebra. To evaluate the limit at infinity of any rational function, we first divide both the numerator and denominator by the highest power of x that occurs in the denominator. (We can assume that $x \neq 0$ because we're interested in only large values of x.) In this case the highest power of x in the denominator is x^2, so we have

$$\lim_{x\to\infty}\frac{3x^2-x-2}{5x^2+4x+1}=\lim_{x\to\infty}\frac{\dfrac{3x^2-x-2}{x^2}}{\dfrac{5x^2+4x+1}{x^2}}=\lim_{x\to\infty}\frac{3-\dfrac{1}{x}-\dfrac{2}{x^2}}{5+\dfrac{4}{x}+\dfrac{1}{x^2}}=\frac{\lim\limits_{x\to\infty}\left(3-\dfrac{1}{x}-\dfrac{2}{x^2}\right)}{\lim\limits_{x\to\infty}\left(5+\dfrac{4}{x}+\dfrac{1}{x^2}\right)}$$

$$=\frac{\lim\limits_{x\to\infty}3-\lim\limits_{x\to\infty}\dfrac{1}{x}-2\lim\limits_{x\to\infty}\dfrac{1}{x^2}}{\lim\limits_{x\to\infty}5+4\lim\limits_{x\to\infty}\dfrac{1}{x}+\lim\limits_{x\to\infty}\dfrac{1}{x^2}}=\frac{3-0-0}{5+0+0}=\frac{3}{5}.$$

(d) Because both $\sqrt{x^2+1}$ and x are large when x is large, it's difficult to see what happens to their difference, so we use algebra to rewrite the function. We first multiply numerator and denominator by the conjugate radical:

$$\lim_{x\to\infty}(\sqrt{x^2+1}-x)=\lim_{x\to\infty}(\sqrt{x^2+1}-x)\cdot\frac{\sqrt{x^2+1}+x}{\sqrt{x^2+1}+x}$$

$$=\lim_{x\to\infty}\frac{(x^2+1)-x^2}{\sqrt{x^2+1}+x}=\lim_{x\to\infty}\frac{1}{\sqrt{x^2+1}+x}.$$

Notice that the denominator of this last expression ($\sqrt{x^2+1}+x$) becomes large as $x\to\infty$ (it's bigger than x). So

$$\lim_{x\to\infty}(\sqrt{x^2+1}-x)=\lim_{x\to\infty}\frac{1}{\sqrt{x^2+1}+x}=0.$$

(e) Divide by the highest power of x in the denominator.

$$\lim_{x\to\infty}\frac{\sqrt{x}+x^2}{2x-x^2}=\lim_{x\to\infty}\frac{(\sqrt{x}+x^2)/x^2}{(2x-x^2)/x^2}=\lim_{x\to\infty}\frac{1/x^{3/2}+1}{2/x-1}=\frac{0+1}{0-1}=-1.$$

(f) Divide by the highest power of x in the denominator.

$$\lim_{x\to\infty}\frac{x^4-3x^2+x}{x^3-x+2}=\lim_{x\to\infty}\frac{(x^4-3x^2+x)/x^3}{(x^3-x+2)/x^3}=\lim_{x\to\infty}\frac{x-3/x+1/x^2}{1-1/x^2+2/x^3}=\infty,$$

because the numerator increases without bound and the denominator approaches 1 as $x \to \infty$.

Problem 4.9 Evaluate the following limits.

(a) $\displaystyle\lim_{\theta\to 0}\frac{\sin 5\theta}{\theta}$

(b) $\displaystyle\lim_{\theta\to 0}\frac{\sin\theta}{2\theta}$

(c) $\displaystyle\lim_{x\to\infty}\sin\frac{1}{x}$

(d) $\displaystyle\lim_{x\to\infty}x\sin\frac{1}{x}$

(e) $\displaystyle\lim_{x\to\infty}\cos\frac{500}{x}$

(f) $\displaystyle\lim_{x\to 0}\frac{\sin^2 x}{3x^2}$

(g) $\displaystyle\lim_{x\to 0}\frac{\sin 2x}{\sin 3x}$

Solution

(a) $\displaystyle\lim_{\theta\to 0}\frac{\sin 5\theta}{\theta}=\lim_{\theta\to 0}\left(5\cdot\frac{\sin 5\theta}{5\theta}\right)=5\lim_{\theta\to 0}\frac{\sin 5\theta}{5\theta}=5\cdot 1=5.$

(b) $\displaystyle\lim_{\theta\to 0}\frac{\sin\theta}{2\theta}=\frac{1}{2}\lim_{\theta\to 0}\frac{\sin\theta}{\theta}=\frac{1}{2}\cdot 1=\frac{1}{2}.$

(c) As $x \to \infty$, $1/x \to 0$ and $\sin(1/x) \to \sin 0 = 0$. Hence

$$\lim_{x\to\infty}\sin\frac{1}{x}=0.$$

(d) $\displaystyle\lim_{x\to\infty}x\sin\frac{1}{x}=\lim_{x\to\infty}\frac{\sin\frac{1}{x}}{\frac{1}{x}}=1.$

(e) $\displaystyle\lim_{x\to\infty}\cos\frac{500}{x}=\cos 0=1.$

(f) $\displaystyle\lim_{x\to 0}\frac{\sin^2 x}{3x^2}=\frac{1}{3}\lim_{x\to 0}\left(\frac{\sin x}{x}\right)^2=\frac{1}{3}\cdot 1^2=\frac{1}{3}.$

(g) $\displaystyle\lim_{x\to 0}\frac{\sin 2x}{\sin 3x}=\lim_{x\to 0}\left(\frac{2}{3}\cdot\frac{3x}{2x}\cdot\frac{\sin 2x}{\sin 3x}\right)=\frac{2}{3}\lim_{x\to 0}\left(\frac{3x}{\sin 3x}\cdot\frac{\sin 2x}{2x}\right)=\frac{2}{3}\cdot 1\cdot 1=\frac{2}{3}.$

Problem 4.10 Verify the limit (3) in Problem 4.5 numerically by using a calculator to construct a table of values of $(1 - \cos\theta)/\theta$ corresponding to the same θ's used in the problem.

Solution

θ	$(1 - \cos\theta)/\theta$	θ	$(1 - \cos\theta)/\theta$
1	0.45969769	0.1	0.04995835
0.5	0.24483488	0.05	0.02499479
0.4	0.19734751	0.01	0.00499996
0.3	0.14887837	0.005	0.00249999
0.2	0.09966711	0.001	0.00050000

Problem 4.11 The limit (2) in Problem 4.5 says that $(\sin\theta)/\theta \cong 1$ or $\sin\theta \cong \theta$ for small θ. Test this approximation by using a calculator to find the value of $\sin\theta$ for (a) $\theta = 0.1$; (b) $\theta = 0.01$; (c) $\theta = 0.001$. Give a geometric explanation for the fact that each $\sin\theta$ is slightly less than its corresponding θ.

Solution

(a) $\sin 0.1 \cong 0.09983342$

(b) $\sin 0.01 \cong 0.00999983$

(c) $\sin 0.001 \cong 0.00099999983$

In the following figure, the arc length 2θ is strictly greater than the chord length $2\sin\theta$ for small positive θ. Hence for such θ, $2\theta > 2\sin\theta$, or $\theta > \sin\theta$.

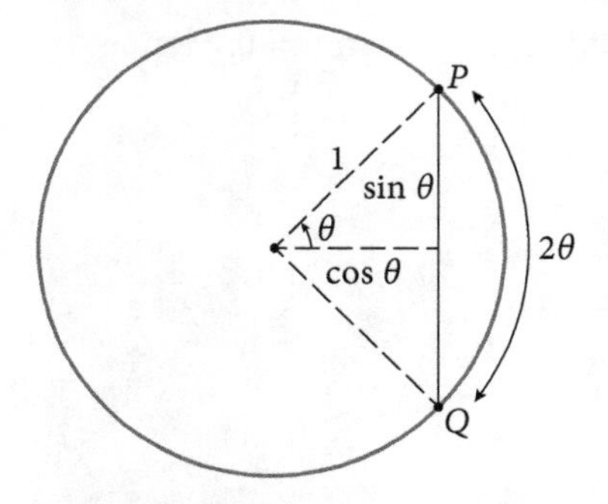

Problem 4.12 Consider the limit $\lim_{x \to 0+} x^x$.

(a) Use a calculator to construct a table of values of x^x for $x = 1, 0.9, 0.8, 0.7, 0.6, 0.5, 0.4, 0.3, 0.2, 0.1, 0.05, 0.01, 0.005, 0.001$. Use this evidence to form a conjecture about the value of the limit.

(b) Use the information in (a) to sketch the graph of $y = x^x$ for $0 < x < 1$. Estimate the location of the lowest point.

Solution

(a)

x	x^x
1	1
0.9	0.9095
0.8	0.8365
0.7	0.7991
0.6	0.7360
0.7	0.7071
0.4	0.6931
0.3	0.6968
0.2	0.7248
0.1	0.7943
0.05	0.8609
0.01	0.9550
0.005	0.9739
0.001	0.9931

From the table, it appears that $\lim_{x \to 0^+} x^x = 1$.

(b) The lowest point is approximately (0.3679, 0.6922), where each coordinate is rounded to four decimal places.

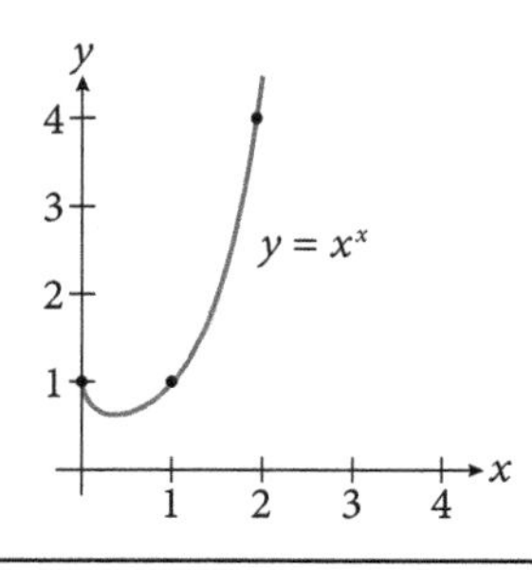

5 Continuous Functions

Review of Continuous Functions and Related Theorems

The first part of this chapter serves as a mathematical reference for the problems in the latter part. Feel free to skip to the problems (page 55) if you're already familiar with continuous functions, the Mean Value Theorem, the Extreme Value Theorem, and the Intermediate Value Theorem.

Continuous Functions

In everyday speech a "continuous" process is one that proceeds without gaps or interruptions or sudden changes. Roughly speaking, a function $y = f(x)$ is continuous if it displays similar behavior; that is, if a small change in x produces a small change in the corresponding value $f(x)$. The function shown in the following figure is continuous at the point a because $f(x)$ is close to $f(a)$ when x is close to a, or more precisely, because $f(x)$ can be made as close as we please to $f(a)$ by taking x sufficiently close to a.

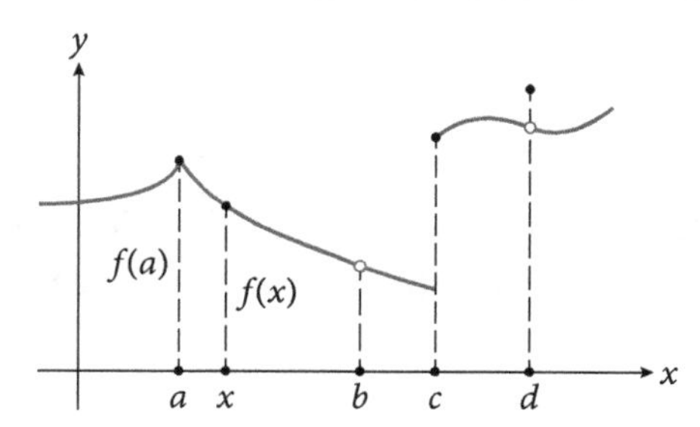

In the language of limits this says that

(1) $\lim_{x \to a} f(x) = f(a).$

Equation (1) is the definition of the statement that $f(x)$ is **continuous** at a. Note that the continuity of $f(x)$ at a has three requirements:

a must be in the domain of $f(x)$, so that $f(a)$ exists;

$f(x)$ must have a limit as x approaches a; and

this limit must equal $f(a)$.

We can understand these ideas more clearly by examining the figure above, in which the function is discontinuous in different ways at the points b, c, and d: at the point b, $\lim_{x\to b} f(x)$ exists but $f(b)$ does not; at c, $f(c)$ exists but $\lim_{x\to c} f(x)$ does not; and at d, $f(d)$ and $\lim_{x\to d} f(x)$ both exist but have different values. The graph of this function therefore has "gaps" or "holes" of three different kinds.

The definition given here tells us what it means for a function to be continuous at a particular point in its domain. A function is called a **continuous function** if it is continuous at every point in its domain. In particular, by the properties of limits we can easily see this to be true for all polynomials and rational functions; and by inspecting their graphs, we see that the functions $\sqrt{x}$, $\sin x$, and $\cos x$ are also continuous. Our interest lies especially in functions that are continuous on closed intervals. These functions are often described as those whose graphs can be drawn without lifting the pencil from the paper.

With a slight change of notation, we can express the continuity of our function $f(x)$ at a point x (instead of a) in either of the equivalent forms

$$\lim_{\Delta x\to 0} f(x+\Delta x) = f(x)$$

or

$$\lim_{\Delta x\to 0} [f(x+\Delta x) - f(x)] = 0;$$

and if we let $\Delta y = f(x + \Delta x) - f(x)$, then this condition becomes

$$\lim_{\Delta x\to 0} \Delta y = 0.$$

This reformulation makes it possible to give a very short proof that *a function that is differentiable at a point is continuous at that point.* The proof occupies only a single line:

$$\lim_{\Delta x\to 0} \Delta y = \lim_{\Delta x\to 0} \frac{\Delta y}{\Delta x}\cdot\Delta x = \left[\lim_{\Delta x\to 0} \frac{\Delta y}{\Delta x}\right]\left[\lim_{\Delta x\to 0} \Delta x\right] = \frac{dy}{dx}\cdot 0 = 0.$$

The converse of this statement isn't true because a function can easily be continuous at a point without being differentiable there (for example, see the point a in the figure above).

The Mean Value Theorem (MVT)

The Mean Value Theorem Let $y = f(x)$ be a function with the following two properties:

$f(x)$ is continuous on the closed interval $[a, b]$; and

$f(x)$ is differentiable on the open interval (a, b).

Then there exists at least one point c in the open interval (a, b) such that

$$(2) \quad f'(c) = \frac{f(b)-f(a)}{b-a},$$

or equivalently,

(3) $f(b) - f(a) = f'(c)(b - a).$

We can see that this statement is reasonable by looking at its geometric meaning as shown in the following figure.

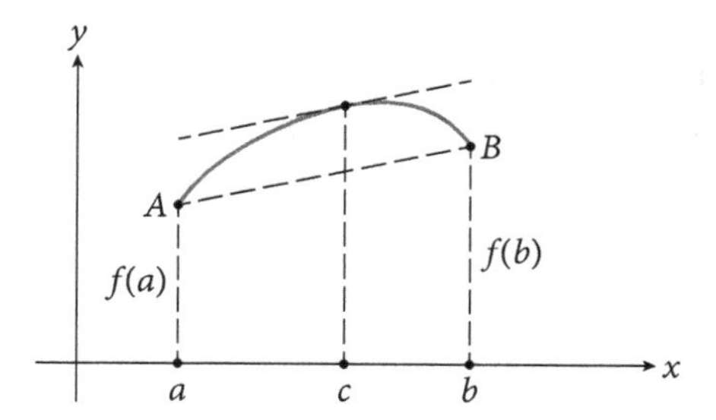

The right side of equation (2) is the slope of the chord joining the endpoints A and B of the graph, and the left side is the slope of the tangent line at the point on the graph corresponding to $x = c$; and the MVT says that for at least one intermediate point on the graph the tangent is parallel to the chord. In the following figure there are two such points, corresponding to $x = c_1$ and $x = c_2$. But this is acceptable because the phrase "at least one point c" allows for the possibility of two such points, or three, or any number ≥ 1.

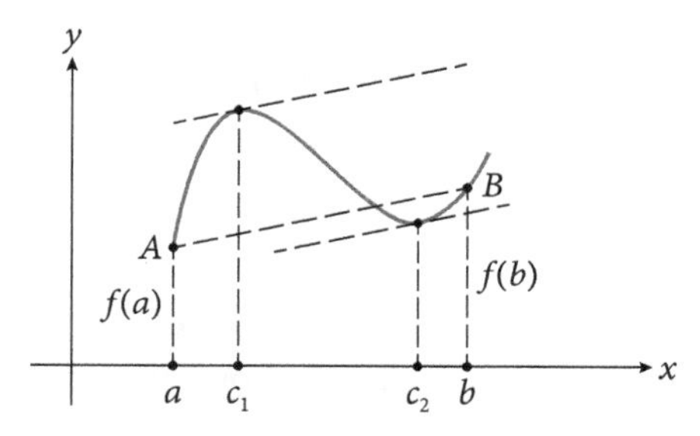

The conclusion of the MVT depends crucially on its hypotheses because this conclusion doesn't follow if the hypotheses are weakened even slightly. We see this by considering the example of the function $y = |x|$ defined on the closed interval $[-1, 1]$. This function, shown in the following figure, is continuous on the closed interval $[-1, 1]$ and is differentiable on the open interval $(-1, 1)$, except at the single point $x = 0$, where the derivative does not exist. The conclusion fails, because the chord joining A and B is horizontal and clearly the graph has no horizontal tangent.

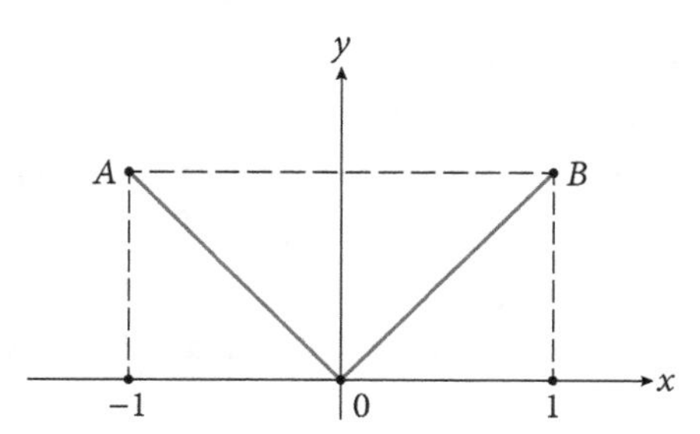

Consequences of the MVT To understand the significance of the Mean Value Theorem, we briefly and informally consider three simple consequences. In each case we have a property of the derivative that implies a property of the function, and the MVT is the link between the two properties.

1. If $f'(x) > 0$ on an interval, then $f(x)$ is increasing on that interval ["increasing" means that $a < b$ implies $f(a) < f(b)$]. Our geometric intuition assures us that this is true, because $f'(x) > 0$ means that the tangent points up to the right everywhere, as illustrated in the following figure. For a more explicit argument based on the MVT, note that in this situation the right side of (3) is positive, so the left side is also positive, and this means that $f(a) < f(b)$.

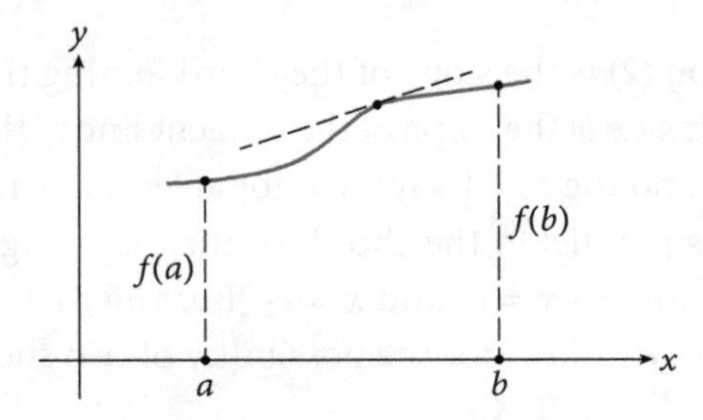

2. Similarly, if $f'(x) < 0$ on an interval, then $f(x)$ is decreasing on that interval ["decreasing" means that $a < b$ implies $f(a) > f(b)$].

3. If $f'(x) = 0$ on an interval, then $f(x)$ is constant on that interval. To show this we assume the contrary, namely, that the function isn't constant. Then there exist two points a and b with $a < b$ at which the function has different values $f(a)$ and $f(b)$. But this implies that the left side of (3) isn't equal to 0, whereas the right side must equal 0. This contradiction shows that our assumption—that the function is not constant—can't be true.

The Extreme Value Theorem (EVT) and Fermat's Theorem

The Extreme Value Theorem If $y = f(x)$ is a function that is defined and continuous on a closed interval $[a, b]$, then this function attains both a maximum value and a minimum value at points of the interval; that is, there exist points c and d in $[a, b]$ such that $f(c) \geq f(x) \geq f(d)$ for all x in $[a, b]$.

The EVT is so-called because maximum values and minimum values are known collectively as **extreme values**. Informally, this theorem asserts that the graph of a continuous function on a closed interval always has both a high point and a low point. If we think of the graph as drawn by moving a pencil across the paper from the point A to the point B (see the following figure), then the statement is so clearly true that we wonder how it could ever be in doubt. It's difficult to prove rigorously, however, because it depends on a subtle property of the real line (completeness, meaning that no points are "missing" from the line) that's normally discussed only in advanced courses.

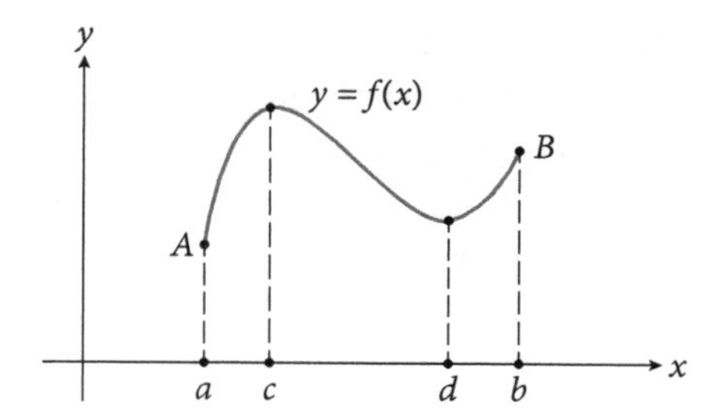

Also, just as in the case of the Mean Value Theorem, the conclusion here depends crucially on the hypotheses that the function is *continuous* and the interval is *closed*. For example, the function in the following figure is continuous on the interval $[0, 1)$, but this interval isn't closed because it lacks the right endpoint. We see that this function attains *no* maximum value at any point of $[0, 1)$ because the only possible candidate for a maximum value is 1 at $x = 1$, but $f(1)$ is not defined.

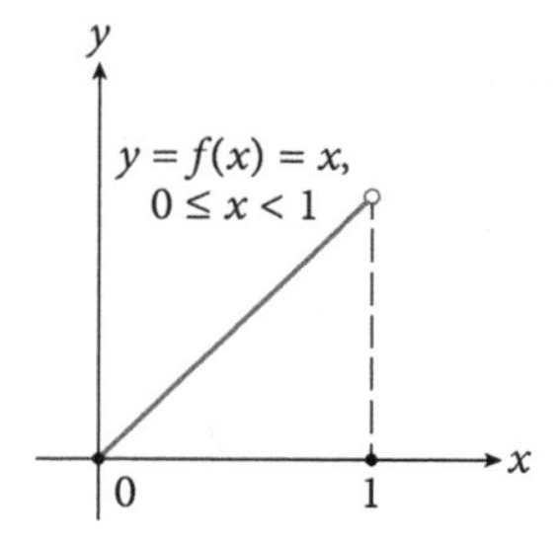

On the other hand, in the following figure the interval $[0, 2]$ is closed and the function is continuous at every point of this interval except for the single point $x = 0$, and again the function attains no maximum value at any point of the interval.

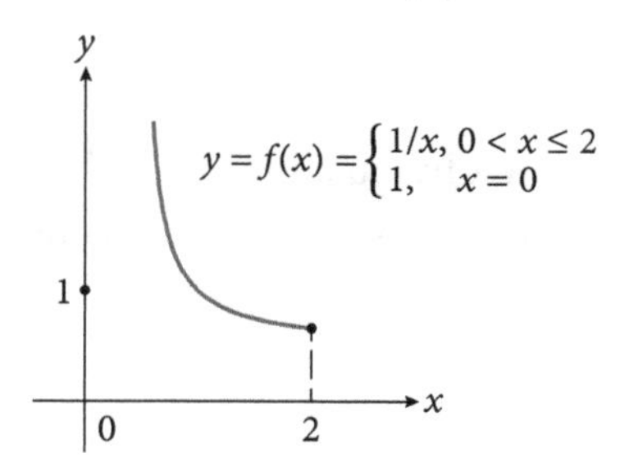

Another important fact about extreme values is **Fermat's theorem**: if a continuous function $f(x)$ on a closed interval $[a, b]$ attains its maximum or minimum value at an interior point c of $[a, b]$, and if $f(x)$ is differentiable at c, then $f'(c) = 0$. Calculus problems often require us to locate extreme values of continuous functions on closed intervals. Fermat's theorem tells us to seek such points either at the endpoints of the interval or at those interior points where $f'(x) = 0$ or $f'(x)$ does not exist.

The Intermediate Value Theorem (IVT)

The Intermediate Value Theorem If $y = f(x)$ is a function that is defined and continuous on a closed interval $[a, b]$, then this function assumes every value between $f(a)$ and $f(b)$; that is, if K is any number strictly between $f(a)$ and $f(b)$, then there exists at least one point c in (a, b) such that $f(c) = K$.

In the language of graphs, every horizontal line of height K intersects the graph of $y = f(x)$ if K is between $f(a)$ and $f(b)$.

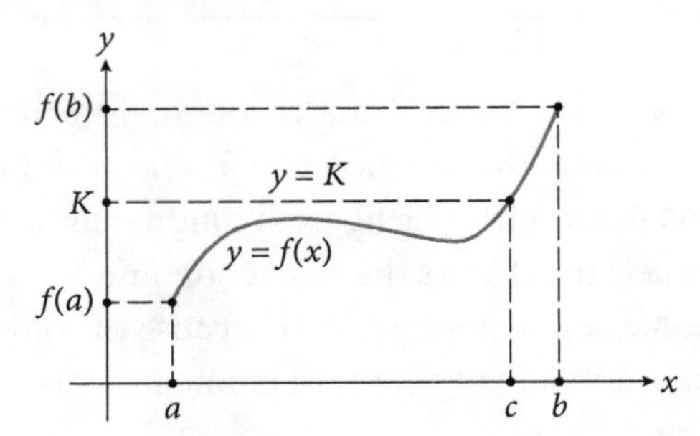

The most striking form of the IVT says that if $y = f(x)$ is continuous on $[a, b]$ and $f(a)$ and $f(b)$ have opposite signs, then $f(c) = 0$ for at least one point c in (a, b). In other words, the graph can't get from one side of the x-axis to the other without actually crossing this axis, as illustrated in the following figure.

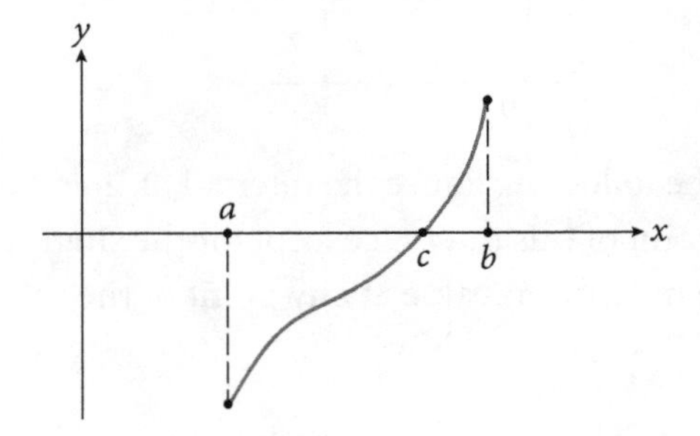

This statement might seem to be obvious, but it is false if the function fails to be continuous at even a single point. We see this by considering the function defined on $[0, 2]$ by

$$y = f(x) = \begin{cases} -1 & \text{if } 0 \le x < 1, \\ 1 & \text{if } 1 \le x \le 2. \end{cases}$$

It's clear that for this function, shown in the following figure, we have $f(0) < 0$ and $f(2) > 0$, and yet—because of the discontinuity at the single point $x = 1$—there does not exist any point c in $(0, 2)$ for which $f(c) = 0$.

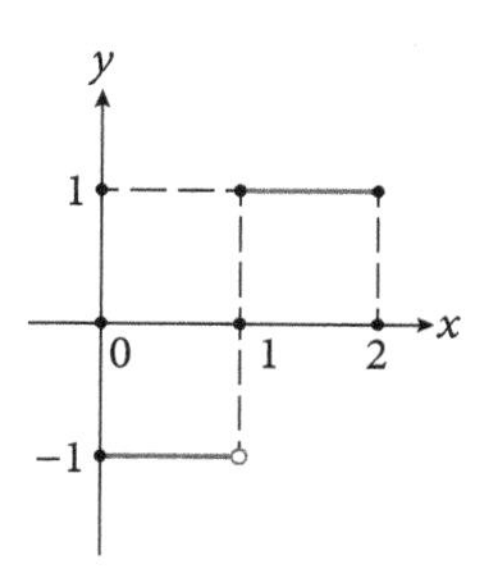

The practical significance of the IVT can best be understood by means of an example. We observe that the equation

(4) $x^3 + 2x - 4 = 0$

isn't easy to solve by factoring, because the left side has no obvious factors. However, the continuous function $f(x) = x^3 + 2x - 4$ is negative at $x = 1$ and positive at $x = 2$ [$f(1) = -1$ and $f(2) = 8$]. The IVT therefore guarantees that $f(x)$ has a zero at some point in (1, 2), so equation (4) has a solution in this interval. Further, $f'(x) = 3x^2 + 2 > 0$ for all x, so $f(x)$ has only one zero and (4) has only one solution. This follows from the fact that if there were two zeros, then the Mean Value Theorem would imply that $f'(c) = 0$ for some intermediate point c, which can't happen.

Problems

Problem 5.1 Find the points of discontinuity of the following functions.

(a) $\dfrac{x}{x^2+1}$

(b) $\dfrac{x}{x^2-1}$

(c) $\dfrac{x^2-1}{x-1}$

(d) $\sqrt{x}$

(e) $\dfrac{1}{\sqrt{x}}$

(f) $\sqrt{x^2}$

(g) $\dfrac{1}{x^2+x-12}$

(h) $\dfrac{1}{x^2+4x+5}$

Solution

(a) No points of discontinuity.

(b) Discontinuous at 1 and −1 because it's undefined there (the denominator becomes zero).

(c) Discontinuous at 1 because it's undefined there.
(d) Discontinuous for all $x < 0$ because it's undefined there.
(e) Discontinuous for all $x \le 0$ because it's undefined there.
(f) No points of discontinuity.
(g) Discontinuous at 3 and -4 because it's undefined there (the denominator becomes 0).
(h) No points of discontinuity because $x^2 + 4x + 5 = (x + 2)^2 + 1 > 0$ for all x.

Problem 5.2 Verify that the function $f(x)$ satisfies the hypotheses of the Mean Value Theorem on the given interval, and find all points c whose existence is guaranteed by the theorem.

(a) $f(x) = x^2 + 1$, $[1, 2]$
(b) $f(x) = x^2 - 4x + 6$, $[2, 4]$
(c) $f(x) = \sqrt{x+1}$, $[0, 3]$
(d) $f(x) = 1/x$, $[1/2, 2]$

Solution (a) $f(x) = x^2 + 1$ is a polynomial, hence continuous on $[1, 2]$. Because $f'(x) = 2x$, $f(x)$ is differentiable on $(1, 2)$. So the MVT applies with $a = 1$, $b = 2$, and there exists c in $(1, 2)$ such that $2c = f'(c) = [f(2) - f(1)]/(2 - 1) = (5 - 2)/1 = 3$. Thus $c = 3/2$ is the only such point in $(1, 2)$.

(b) For the same reasons as in part (a), the hypotheses of the MVT are satisfied with $f'(x) = (x^2 - 4x + 6)' = 2x - 4$. There exists c in $(2, 4)$ such that $2c - 4 = f'(c) = [f(4) - f(2)]/(4 - 2) = (6 - 2)/(4 - 2) = 2$. So $c = 3$ is the only such point.

(c) $f(x) = \sqrt{x+1}$ is defined and continuous on $[0, 3]$. By rationalizing the numerator of the difference quotient, we find that $f(x)$ has the derivative

$$f'(x) = \lim_{\Delta x \to 0} \frac{\sqrt{x+\Delta x+1} - \sqrt{x+1}}{\Delta x}$$

$$= \lim_{\Delta x \to 0} \frac{1}{\sqrt{x+\Delta x+1} + \sqrt{x+1}} = \frac{1}{2\sqrt{x+1}},$$

which is defined on $(0, 3)$. Hence the MVT applies, and there exists c in $(0, 3)$ such that

$$\frac{1}{2\sqrt{c+1}} = f'(c) = \frac{f(3) - f(0)}{3-0} = \frac{2-1}{3-0} = \frac{1}{3}.$$

So $\sqrt{c+1} = 3/2$ and squaring both sides yields $c + 1 = 9/4$. Thus $c = 5/4$ is the only such point.

(d) $f(x) = 1/x$ is defined and continuous on $[1/2, 2]$. By Problem 2.5, $f'(x) = -1/x^2$, and $f(x)$ is differentiable on $(1/2, 2)$. So the MVT applies and there exists c in $(1/2, 2)$ such that

$$-\frac{1}{c^2} = f'(c) = \frac{f(2) - f(1/2)}{2-(1/2)} = \frac{(1/2)-2}{2-(1/2)} = -1$$

Then $c^2 = 1$, and $c = \pm 1$. Only the point $c = 1$ lies in $(1/2, 2)$, and hence $c = 1$ is the only point guaranteed by the MVT.

Problem 5.3 If $f(x) = 1/x$ and $g(x) = 1/x + x/|x|$, then show that these functions have identical derivatives, so that $[f(x) - g(x)]' = 0$. Their difference isn't constant, however. Explain how this is possible in view of consequence 3 of the Mean Value Theorem.

Solution By Problem 2.5, the derivative of $f(x) = 1/x$ is $f'(x) = -1/x^2$. For $g(x) = 1/x + x/|x|$, note that $g(x) = 1/x + 1$ for $x > 0$ and $g(x) = 1/x - 1$ for $x < 0$. At $x = 0$, $g(x)$ is undefined. As we saw in several problems in Chapter 2, the constant term of a function vanishes when the derivative is computed (it subtracts out in the difference quotient). Hence $g'(x) = -1/x^2$, the derivative of $f(x) = 1/x$. That is, $f(x)$ and $g(x)$ have the same derivative, although $f(x) - g(x)$ isn't constant, this difference being equal to -1 or 1, depending on the sign of x. This fact isn't a violation of the MVT, however, because over any interval that includes $x = 0$, $f(x) - g(x)$ isn't defined throughout the interval, as the MVT demands. On the other hand, if an interval doesn't include $x = 0$, then $f(x) - g(x)$ will be constant, as the theorem requires.

Problem 5.4 If $f'(x) = c$, a constant, for all x, then show that $f(x) = cx + d$ for some constant d.

Solution In the same manner as Problem 5.3, we have $(f(x) - cx)' = f'(x) - (cx)' = f'(x) - c = f'(x) - f(x) = 0$, for all x. Thus $f(x) - cx$ is equal to a constant, say d, for all x, and $f(x) = cx + d$ as claimed.

Problem 5.5 For each of the given intervals, find the maximum value of $\sin x$ on that interval, and also find the value of x at which it occurs.

(a) $[0, \pi/6]$

(b) $[0, \pi/4]$

(c) $[0, \pi]$

Solution Inspecting the following figure reveals that as $\theta = x$ increases from 0 to $\pi/2$ radians, $\sin x$ increases from 0 to 1 and as $\theta = x$ increases from $\pi/2$ to π radians, $\sin x$ decreases from 1 to 0.

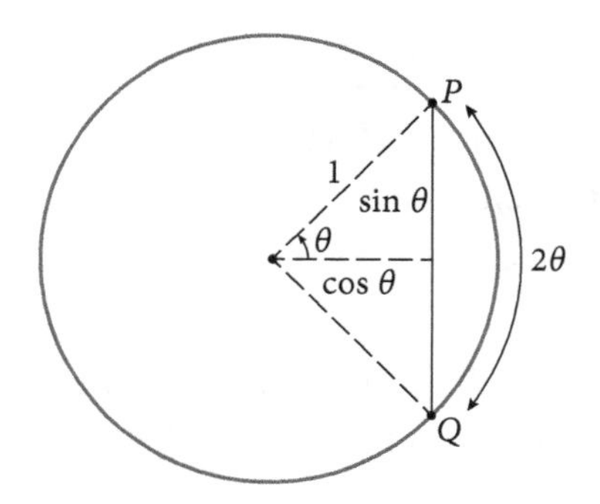

(a) By the above remarks, $\sin x$ takes on the maximum value on $[0, \pi/6]$ when $x = \pi/6$, where $\sin x = \pi/6 = 1/2$.

(b) As in part (a), $\sin x$ takes on the maximum value on $[0, \pi/4]$ when $x = \pi/4$, where $\sin x = \sin \pi/4 = \sqrt{2}/2$.

(c) By the above remarks, $\sin x$ takes on the maximum value of 1 on $[0, \pi]$, when $x = \pi/2$.

Problem 5.6 Does the rational function $\dfrac{x^3 - x^5}{1 + 9x^4 + 5x^6}$ have

(a) a maximum value on [5, 8]?
(b) a minimum value on [5, 8]?

Solution (a), (b) Yes. This function is defined and continuous on [5, 8] because the denominator $1 + 9x^4 + 5x^6$ is nonzero on [5, 8]. By the Extreme Value Theorem, the function has a maximum and minimum value on [5, 8].

Problem 5.7 Does the function x^4 have a minimum value on the following intervals? If so, where?

(a) [–3, 5]
(b) (–4, 2)
(c) (2, 3)
(d) (–1, 5)

Solution (a) Yes. x^4 is continuous on [–3, 5], so by the Extreme Value Theorem x^4 has a minimum value on [–3, 5]. Because $(x^4)' = 4x^3 = 0$ for $x = 0$, Fermat's theorem says that the minimum must occur at $x = 0$ or the endpoints –3 and 5. Checking these three values, we find that the minimum value occurs at $x = 0$.

(b) Yes. $x^4 = 0$ only when $x = 0$, and $x^4 > 0$ for $x \neq 0$. Hence x^4 attains its minimum value on (–4, 2) at $x = 0$.

(c) No. Because $(x^4)' = 4x^3 \neq 0$ for x in (2, 3), x^4 has no minimum value on (2, 3).

(d) Yes. By the same reasoning as part (b), x^4 attains its minimum value on (–1, 5) at $x = 0$.

Problem 5.8 Does the function $4 + x^2$ have

(a) a maximum value on (–2, 2)?
(b) a minimum value on (–2, 2)?

If so, where?

Solution (a) No. Because $(4 + x^2)' = 2x = 0$ only when $x = 0$, and $4 + 1^2 = 5 > 4 + 0^2 = 4$, for example, a maximum value on (–2, 2) doesn't occur at $x = 0$, hence at no point in (–2, 2).

(b) Yes. Because $4 + x^2 \geq 4$, with equality occurring at only $x = 0$, the minimum value on $(-2, 2)$ occurs at $x = 0$.

Problem 5.9 Find the maximum and minimum values attained by the given function on the given interval.
(a) $1/x$, $(0, 1]$
(b) $1 - x^2$, $[0, 1)$
(c) $1 - x^2$, $[-3, -2]$
(d) $2 + |2x - 3|$, $(0, 2)$

Solution (a) Because $(1/x)' = -1/x^2 < 0$ for all x, no maximum or minimum value occurs on $(0, 1)$. By consequence 2 of the Mean Value Theorem, $1/x$ is decreasing on $(0, 1)$. So a minimum value of $1/1 = 1$ occurs on $(0, 1]$ at $x = 1$, and no maximum value occurs on $(0, 1]$.

(b) Because $(1 - x^2)' = -2x < 0$ for x in $(0, 1)$, no maximum or minimum value occurs on $(0, 1)$. By consequence 2 of the MVT, $1 - x^2$ is decreasing on $(0, 1)$. So a maximum value of $1 - 0^2 = 1$ occurs on $[0, 1)$ at $x = 0$, and no maximum value occurs on $[0, 1)$.

(c) Because $(1 - x^2)' = -2x > 0$ for x in $(-3, -2)$, all extreme values of $1 - x^2$ on $[-3, -2]$ occur at $x = -3$ and $x = -2$. At $x = -3$, $1 - x^2 = 1 - (-3)^2 = -8$ while at $x = -2$, $1 - x^2 = 1 - (-2)^2 = -3$. So the maximum and minimum values of $1 - x^2$ on $[-3, -2]$ are -3 and -8, respectively.

(d) Let $f(x) = 2 + |2x - 3|$. For $x > 3/2$, $f(x) = 2 + (2x + 3) = 2x + 5$ and $f'(x) = 2$. For $x < 3/2$, $f(x) = 2 - (2x + 3) = -2x - 1$, and $f'(x) = -2$. At $x = 3/2$, $f'(x)$ does not exist. So any extreme value can occur at only $x = 3/2$, where $f(3/2) = 2 + |2 \cdot 3/2 - 3| = 2$. By consequences 1 and 2 of the MVT, $f(x)$ is increasing for $x > 3/2$ and decreasing for $x < 3/2$. So the maximum value of $f(x)$ on $(0, 2)$ is 2, and no minimum value occurs.

Problem 5.10 Apply the Intermediate Value Theorem to show that the given equation has a solution in the given interval.
(a) $x^4 + 3x - 5 = 0$, $[1, 2]$
(b) $x^6 - 3x + 1 = 0$, $[-1, 1]$

Solution (a) Let $f(x) = x^4 + 3x - 5$. Then $f(1) = 1^4 + 3(1) - 5 = -1 < 0$ and $f(2) = 2^4 + 3(2) - 5 = 17 > 0$. By the IVT, $f(c) = 0$ for some point c in $(1, 2)$ because $f(1) < 0 < f(2)$.

(b) Let $f(x) = x^6 - 3x + 1$. Then $f(-1) = (-1)^6 - 3(-1) + 1 = 5 > 0$ and $f(1) = 1^6 - 3(1) + 1 = -1 < 0$. By the IVT, $f(c) = 0$ for some point c in $(-1, 1)$ because $f(-1) > 0 > f(1)$.

Problem 5.11 If $p(x)$ is a polynomial of odd degree, then show that the equation $p(x) = 0$ has at least one solution.

Solution Let $p(x) = a_n x^n + a_{n-1}x^{n-1} + \cdots + a_0$, where $n > 0$ is odd. Suppose that $a_n > 0$; the case when $a_n < 0$ is similar. For sufficiently large positive x, the term $a_n x^n$ is positive and dominates the lower-degree terms of $p(x)$ in absolute value; hence $p(x)$ is positive. Similarly, for negative x sufficiently large in absolute value, $p(x)$ is negative because n is odd and $a_n > 0$. So by the Intermediate Value Theorem, $p(x)$ must be 0 at some point between these two extremes.

Problem 5.12 If $f(x)$ and $g(x)$ are continuous on $[a, b]$, and if $f(a) < g(a)$ and $f(b) > g(b)$, then show that the equation $f(x) = g(x)$ has at least one solution in (a, b).

Solution Let $h(x) = f(x) - g(x)$. Then $h(x)$ is continuous on $[a, b]$ because so are $f(x)$ and $g(x)$, and $h(a) = f(a) - g(a) < 0$ while $h(b) = f(b) - g(b) > 0$. By the Intermediate Value Theorem, $h(c) = 0$ for some point c in (a, b). That is, $h(c) = f(c) - g(c) = 0$, and $f(c) = g(c)$.

Problem 5.13 Let $y = f(x)$ be a continuous function defined on the closed interval $[0, b]$ with the property that $0 < f(x) < b$ for all x in $[0, b]$. Show that there exists a point c in $(0, b)$ with the property that $f(c) = c$. Hint: Consider the function $g(x) = f(x) - x$.

Solution Because $f(x)$ is continuous on $[0, b]$, so is $g(x) = f(x) - x$. As $0 < f(x) < b$ for all x in $[0, b]$, $g(0) = f(0) - 0 = f(0) > 0$ and $g(b) = f(b) - b < 0$. By the Intermediate Value Theorem, $g(c) = f(c) - c = 0$ for some point c in $(0, b)$, and $f(c) = c$.

Problem 5.14 For what values of x is f continuous?

$$f(x) = \begin{cases} 0 & \text{if } x \text{ is rational,} \\ 1 & \text{if } x \text{ is irrational.} \end{cases}$$

Solution $f(x)$ is continuous nowhere. For, given any number a and any $\delta > 0$, the interval $(a - \delta, a + \delta)$ contains both infinitely many rational and infinitely many irrational numbers. Because $f(a) = 0$ or 1, there are infinitely many numbers x with $0 < |x - a| < \delta$ and $|f(x) - f(a)| = 1$. Thus, $\lim_{x \to a} f(x) \neq f(a)$. In fact, $\lim_{x \to a} f(x)$ does not even exist.

Problem 5.15 Does a number exist that is exactly 1 more than its cube?

Solution If such a number exists, then it satisfies the equation $x^3 + 1 = x$ or, equivalently, $x^3 - x + 1 = 0$. Let the left-hand side of this equation be called $f(x)$. Now $f(-2) = -5 < 0$, and $f(-1) = 1 > 0$. Note also that $f(x)$ is a polynomial, and thus continuous. So by the Intermediate Value Theorem, there exists a number c between -2 and -1 such that $f(c) = 0$, so that $c = c^3 + 1$.

Printed in the USA
CPSIA information can be obtained
at www.ICGtesting.com
LVHW081622260924
792254LV00020B/190

9 781937 842437